The Epidemic
of
Health Care
Worker Injury

An Epidemiology

The Epidemic
of
Health Care
Worker Injury

An Epidemiology

William Charney and
Guy Fragala

CRC Press
Boca Raton London New York Washington, D.C.

Library of Congress Cataloging-in-Publication Data

The epidemic of health care worker injury : an epidemiology / edited by William Charney and Guy Fragala.
    p. cm.
    Includes bibliographical references and index.
    ISBN 0-8493-3382-2 (hardcover)
    1. Medical personnel--Health and hygiene.   2. Medical personnel--Wounds and injuries--
Epidemiology. I. Charney, William, 1947-   II. Fragala, Guy.
   RC965.M4H43   1998
   617.1'0088'61--dc21

                        98-36748
                          CIP

No claim to original U.S. Government works
International Standard Book Number 0-8493-3382-2
Library of Congress Card Number   98-36748
Printed in the United States of America   1 2 3 4 5 6 7 8 9 0
Printed on acid-free paper

Preface

This book has been written to reveal the phenomenon of the escalating injury rates to health care workers in the United States.

During the last decade injury rates in the acute care sector have risen by 40%. In the long-term health care sector, the injury rates have doubled during this same time period. In home health care, an industry growing by 19% annually, data are still being collected, though preliminary studies show rising rates, especially in back injury. These rates are on one hand an "epidemic" and on the other hand virtually not known to either the public or the "body politique." This book, *The Epidemic of Health Care Worker Injury: An Epidemiology*, (1) explores the data for the major risk categories; (2) looks at the causes and effects of the health care occupational risk, which has risen to the top of the dubious ladder as being one of the most dangerous occupations in the United States; and (3) examines recommendations and possible solutions.

Chapter 1 is an overview of the data, risk categories, and analysis of causation of health care worker injury. In Chapter 2, the five occupational hazards encountered in nursing and health care are discussed, along with prevention and control. Needlesticks, an "epidemic within an epidemic," occur every 3.9 seconds to a health care worker in the United States. These and other sharp injuries are covered in Chapter 3. Back injury in all health care sectors leads most other industries in reasons for lost workdays; and Chapter 4, authored by Dr. Bernice Owen, a recognized national expert, is devoted to this subject. In Chapters 5 through 10, occupational exposure to tuberculosis (5), hepatitis C (6), latex sensitization (7), violence in the health care workplace (8), nursing home health care worker injury (9), and home health care worker injury (10) are examined; and appropriate recommendations are explored. Results of a research project on occupational injury and illness conducted by the Minnesota Nurses Association are cited in Appendix 1. Hospital injury data charts are provided in Appendix 2.

This volume will function as a "one-stop shopping" for researchers and epidemiologists interested in a guide to health care occupational injury.

Editors

William Charney is an industrial hygienist who specializes in hospital systems. For 5 years Mr. Charney was the Director of the Industrial Hygiene and Occupational Health and Safety department at the Jewish General Hospital, a 600-bed tertiary, teaching, and research hospital in Montreal, Quebec. He is the past Director of Environmental Health at San Francisco General Hospital, San Francisco, CA. Mr. Charney is a member of the American Conference of Governmental Industrial Hygienists and the Canadian Society of Safety Engineers, and in 1998 was the recipient of the California State Department of Health Environmental Health Award.

Guy Fragala, Ph.D., PE, CSP, is currently the Director of the Environmental Health and Safety Department at the University of Massachusetts Medical Center in Worcester, MA. He is also a faculty member in the department of Family and Community Medicine within the medical school, serving as the occupational safety specialist for the occupational health program. He established the Medical Center's environmental health and safety and risk management programs and has served as a member of the institution's Total Quality Improvement Council.

Dr. Fragala has a broad background in the field of ergonomics and occupational health and safety including service on the DuPont Corporation's Corporate Environmental Health and Safety staff, and safety engineering work for the Commercial Union Assurance Company. He has served on the faculty of a number of academic institutions including Harvard University, the University of Wisconsin, the University of Southern California, and Worcester Polytechnic Institute. As a faculty member for educational programs offered by the Joint Commission on Accreditation of Health Care Organizations, he worked on the development of a program entitled "Hospital-Wide Quality Assurance: A Multidisciplinary Approach." He has also been a regular contributor to the Joint Commission's Plant, Technology, and Safety Management Update Series. He has served on the board of directors for the American Hospital Association's American Society for Hospital Risk Management, and is on the executive committee of the National Safety Council's health care section. He has delivered numerous presentations on the subject of health care—environmental health and safety, including presentations for the American Hospital Association and participation in national teleconferences on the topic.

Dr. Fragala is a graduate of the University of Massachusetts School of Industrial Engineering and Operations Research and is a registered professional engineer, a certified safety professional, and a certified health care safety professional. He has consulted to a wide range of American industries and authored numerous publications on the subjects of ergonomics and environmental health and safety.

Contributors

Alisa Boyd
Occupational Health Branch
Hazard Evaluation System and
  Information Service
State of California Department of
  Health Services
Berkeley, California

James E. Cone
Occupational Health Branch
Hazard Evaluation System and
  Information Service
State of California Department of
  Health Services
Berkeley, California

Barbara DeBaun
California Pacific Medical Center
San Francisco, California

Elaine El-Askari
Labor Occupational Health Program
School of Public Health
University of California
Berkeley, California

Michael Foley
Safety and Health Assessment and
  Research for Prevention (SHARP)
  Program
Washington State Department of Labor
  and Industries
Olympia, Washington

Kim Hagadone
Occupational Health Branch
Hazard Evaluation System and
  Information Service
State of California Department of
  Health Services
Berkeley, California

John Kalat
Safety and Health Assessment and
  Research for Prevention (SHARP)
  Program
Washington State Department of Labor
  and Industries
Olympia, Washington

Jane Lipscomb
University of Maryland
  School of Nursing
Baltimore, Maryland

Bernice D. Owen
Clinical Science Center
School of Nursing
University of Wisconsin—Madison
Madison, Wisconsin

Bonnie Rogers
Public Health Nursing and the
  Occupational Health Nursing Program
School of Public Health
University of North Carolina at Chapel
  Hill
Chapel Hill, North Carolina

Jon Rosenberg
Infection Control and Healthcare
  Epidemiology
Division of Communicable Disease
  Control
California Department of Health
  Services
Berkeley, California

Barbara Silverstein
Safety and Health Assessment and
  Research for Prevention (SHARP)
  Program
Washington State Department of Labor
  and Industries
Olympia, Washington

George W. Weinert
Risk Management Association
Belmont, Massachusetts

Table of Contents

1 An Epidemic of Health Care Worker Injury*

William Charney

CONTENTS

INTRODUCTION

Health care is fast becoming one of the most dangerous jobs in the United States. Injuries and illnesses are reaching epidemic levels. Two studies report that the increasing trend of health care worker injury has been continuing for more than a decade.[1,2]

    Traditionally, health care worker exposure data are analyzed one category at a time. This approach tends to isolate the researcher from the more global perspective of an industry-wide analysis. The goal of this chapter is to document the epidemiology of an industry with dangerously growing accident and illness rates and to discuss some of the causes creating the upward trend.

TYPES OF INJURIES

The highest incidence of injuries among health care workers is reported among custodial and housekeeping personnel, followed by food service and nutritional

---

* Reprint from W. Charney: An Epidemic of Health Care Worker Injury. *New Solutions.* 1997; 7(3). With permission.

employees, and then nurses and laboratory technicians. Needlestick injuries are the most prevalent type of injury. There are approximately 800,000 needlestick injuries reported each year; combined with a nonreported rate of 40–60%,[3] there is a probability of more than 1.5 million sticks per year. These injuries are followed by strains and sprains, the most expensive injury in health care, more than half of which are related to lifting and twisting while moving patients.[4]

Studies of back-related compensation claims show that nurses and nurses' aides have among the highest claim rates for this injury of any occupation or industry. Jensen[5] documented that patient lifting was the cause of the back injuries. Of nurses who leave the profession, 12% do so because of back injury.[6]

TYPES OF EXPOSURES

Health care workers are an exposed population. A partial roll call of exposures includes the following: infectious diseases from needlesticks, including but not limited to human immunodeficiency virus (HIV) and hepatitis B and C; infectious diseases such as tuberculosis (TB); radiation exposures from radiological equipment and diagnostic and therapeutic radioisotopes; chemical exposures from cold sterilants, chemotherapy drugs, waste anesthetic gases, sterilizer gases (such as ethylene oxide), laser plumes, and aerosolized drugs (such as ribavirin and pentamidine); ergonomic injury, especially back injury; violence, especially in psychiatric hospitals and units; shift work hazards; reproductive hazards; and research laboratorians exposed to research chemicals. Additionally, 8.9% of health care workers developed latex allergy.[7]

A 13-step data review presents a wide-angle view of the dangers of working in the health care industry as follows:

1. The incidence rate of lost work days per 100 full-time workers per year for health care workers in 1994 was 8.7 compared with 6.1 for all service workers combined.[8] In addition, the rates of health care worker injury have been increasing in the last decade. Based on data from the Bureau of Labor Statistics (BLS), injury and illness rates in all health services—including doctors' offices, hospitals, nursing homes, medical labs, and allied health providers—increased from six cases per 100 full-time employees in 1980 to ten cases per 100 full-time employees in 1992.[2]
2. In 1994, the injury and illness rate for the health care sector was 9.4 per 100 full-time workers, greater than in mining (6.3).[9]
3. Occupations with the most injuries and illnesses involving days away from work in 1994 had nursing aides and orderlies with a rate of 101.8 (cases in thousands), surpassed only by laborers (147.3) and truck drivers (163.8).[10]
4. Nursing personnel rank fifth nationally for filing workers' compensation cases; only heavy laborers, such as miscellaneous workers, sanitation workers, warehouse workers, and mechanics, surpass nurses.[11]
5. In 1992, private health services employed 8.5 million workers. In the same year, nearly 700,000 work-related injuries and illnesses were reported.[2]

6. The number of lost workdays per 100 full-time nursing home workers has doubled in the last decade, increasing from 98.2 to 186.9 in 1993.[12]
7. In 1994, nursing aides and orderlies accounted for more than half of all assaulted service workers, attacked primarily by patients or residents who resisted their help or were prone to violence.[10]
8. The incidence of lost workdays for the hospital sector is 2.9 per 100 full-time workers, compared with 2.8 for other private industry.[8]
9. In 1992, injury rates per 100 full-time workers for hospitals was 12 and for nursing homes, 18.6, compared with other private industry, which had 8.9.[2]
10. Health care workers experienced three times the risk of the general population for acquiring hepatitis C.[13]
11. There were 86 fatal occupational injuries in the health services codes for 1994, an approximate average of one health care occupational fatality every 4 days: 31.4% resulted from assaults and violent acts; 50%, from transportation incidents; and 8%, from falls.[14]
12. Home health care aides had an annual low-back injury rate of 15.4 per 100 full-time equivalents (FTEs), significantly higher than the rate for hospital nursing assistants with 5.9 per 100 FTEs. Of the home health care aides, 81% were working alone when injury occurred, compared with 39% of nursing assistants.[15]
13. Data for occupational HIV transmission through 1995 in the United States showed 46 confirmed transmissions and 97 possible transmissions for a total of 143. This is greater than in all of Western Europe.[1] Working in a nursing home is more dangerous for rate of injury than working in a coal mine, a steel mill, a warehouse, or a paper mill. In 1993, the injury and illness rate for nursing homes was an alarming 17.3 per 100 full-time workers, compared with coal mining (10.3), steel mills (14), warehouses (11.9), and paper mills (8.7).[12]

The 13-step review, seen as a totality, points to an industry that is climbing to the top of the dubious ladder of one of the "most dangerous industries" in the United States. Stellman's classic, *Work Is Dangerous to Your Health*, explained how a work environment can compromise health. In a nation where 150 workers die each day as a result of traumatic injury or occupational illness—totaling 50,000 deaths a year—and where there is a disabling injury every 17 sec, the hospital sector is no exception in contributing to these statistics and, as shown earlier, in some cases leads the way.

UPWARD TRENDS

The great question is, "What is causing this epidemic of injury to health care workers and the upward trends in the last decade?" It is not due to the greater number of health care workers in the industry because the data are corrected for this factor. The causes are multiple. Among them are lack of regulation; addition of chemicals, treatments, and medical technology without first testing for occupational risk; lack of appropriate standards based on exposure; inappropriate training based on the level of risk; lack of collective bargaining agreements for health departments; managed care with its principles of downsizing, de-skilling, and increasing acuity of patient conditions; and, finally, lack of occupational health training in both nursing and medical school curricula.

Focusing on *some* of these causes, one at a time, will help illustrate the epidemic proportions of health care worker injury:

REGULATION

A critical analysis of the role of regulatory agencies is necessary to understand the interplay between lack of regulation and increases in injury rates. Federal and state occupational safety and health administrations (OSHAs) and the Joint Commission on Hospital Accreditation Organization (JCAHO) are the main players in the field of occupational health regulation in the hospital sector. The Centers for Disease Control and Prevention (CDC) writes appropriate guidelines for health care organizations, but the guidelines do not have the rule of law. Only 188 inspections in Standard Industry Code (SIC) code 8062—medical hospitals—were accomplished by the federal OSHA in 1995 out of 26,351 inspections industry wide, a total of less than 1%. In the state of California, out of 10,665 inspections by state OSHA industry wide, only 109 inspections occurred in SIC code 8062, or 1%.[16]

OSHA inspectors are undertrained when it comes to the hospital sector. Only now are some state OSHAs beginning to train their personnel to understand specific hospital technologies.

The JCAHO, in its environment of care section, has standards for hazardous waste handling, health and safety committees, emergency preparedness, infection control, written safety programs, and training, but its standards are at best generic and do not deal with the issues of health care worker risk mentioned. In a study released by the JCAHO, more than 40% of 5208 hospitals surveyed from 1986–1988 had less than significant compliance with published safety standards.[17] Even hospitals that have been given high marks in the environment of care section under JCAHO can still be dangerous environments. There are no standards in the JCAHO environment of care section that deal with needlestick prevention or management, the most prevalent risk. No standards exist for ergonomics or back injury, the most expensive risk. There is no mention of ethylene oxide, one of the most toxic chemicals used in hospitals; nothing requires chemical safety plans for hospital laboratories, a lack of which is highly responsible for dangerous laboratory conditions. Again, there is *nothing* about the following: preparation or administrative systems for chemotherapeutic drugs; mandating hospitals to lower their accident and injury rates by even small percentages every 3 years between inspections; requiring scavenging systems for electrocautery or laser plumes; operating room safety for halogenated waste anesthetic gas delivery systems; heat stress in laundries; requiring hospitals to have a safe-handling policy for the many carcinogenic materials that are used; and quality of training programs (only on the number of personnel trained, etc.).

CHEMICALS, TREATMENTS, AND NEW TECHNOLOGY

Health care workers are exposed to a plethora of chemicals, several treatments, and different technologies that have no occupational health standards or have never been tested for occupational effects. Some examples are chemotherapy drugs that include alkylating agents, antimetabolites, and antibiotics. More than 35 drugs are available,[18] none of which had been tested for occupational exposure effects. No developed exposure standards exist for these chemicals.

Inhalation anesthesia agents can cause potential harm if they are not controlled. Three new agents—Isoflurane, Suprane, and Sevoflurane—have been introduced into the workplace with no exposure standards.[19]

Aerosolized medications like ribavirin, which is used to treat pediatric pneumonias, and pentamidine, which is used to treat pneumonia in AIDS patients (the former, a discovered teratogen and the latter, a suspected carcinogen) were introduced into clinical treatment without safeguards for the health care worker delivering the medication.[20]

Electrocautery and laser plumes are carcinogenic and can transmit bioviruses like human papilloma virus (HPV),[21] yet there is no mandated regulation that these plumes must be scavenged and controlled. This laser technology was introduced without testing for downstream occupational health effects. Health care workers handle formaldehyde, which is a carcinogen that has been linked to occupational asthma in the hospital setting.[22] Formaldehyde was introduced into health care before definitive health care worker exposure data were completed. The recommended threshold limit value (TLV) is being lowered continually because of recent toxicity data and now sits at a very low 0.75pm. Glutaraldehyde, a cold sterilant chemical, was introduced into the health care setting without rigorous occupational risk analysis testing. Early studies by the National Institute for Occupational Safety and Health (NIOSH) showed nearly 50% of workers monitored were exposed to levels that exceeded the acceptable ceiling level of 0.2pm.[23] Not until health care workers started to document their health problems did NIOSH conduct studies and write recommendations.

LACK OF APPROPRIATE STANDARDS

Health care workers often are "canaries," exposing their bodies to untested or undertested agents and to a work environment that is underregulated from the occupational safety perspective. There is only one federal OSHA standard, at this writing, specifically for hospitals: the bloodborne pathogens standard. There are other federal and state OSHA standards that can be applied to the hospital sector when problems arise and employees call in complaints. However, despite a long awaited TB standard, which is about to be adopted, but has yet to be promulgated, the fact that there is only one specific regulation written for the health care industry speaks to an ambivalence about protecting health care workers. The bloodborne pathogens standard has been applauded by all concerned as a breakthrough in regulation. Although it mandated hepatitis B vaccine for all health care workers exposed, it is weak about safe-needle technology (only recommending its use), despite the fact that studies now show the effectiveness of using safety devices.[24]

INFECTIOUS DISEASES, THE NEW WORRIES

Occupational transmission risk to health care workers of HIV and hepatitis B is widely documented. Now hepatitis C and TB are the new infectious diseases of growing concern in the health care community. For hepatitis C, there is still no national or cooperative program of surveillance, similar to the HIV effort, to track the disease. One study showed that health care workers are three times more likely to contract the disease than the general population.[13] Another study[25] of dentists showed that this population of health care providers had a 1.75% rate—or 8 out of 456 dentists tested acquired the disease occupationally. The same study showed that oral surgeons had a 10% prevalence rate (4 out of 43), and this population is now considered at high risk. In yet another study, 6% of 50 health care workers with a needlestick seroconverted.[26] The CDC published revised guidelines for following up exposure to hepatitis C virus (HCV); it is mentioned explicitly that there is no postexposure prophylaxis for the disease.[27] TB has reemerged as a definite health threat to health care workers, especially with the development of multi-drug-resistant strain of TB. TB transmission is a recognized risk in health care facilities.[28] The magnitude of the risk varies considerably by type of facility. Several studies have documented higher-than-expected health care worker conversion rates.[29, 30] There are many cases of health care workers acquiring active TB from occupational transmission.[28] The CDC and the others have investigated several outbreaks of multi-drug-resistant TB in hospitals in New York and Florida.[31] Seventeen health care workers were infected in a South Florida clinic.[32] In New York City, where the TB rates have soared since the mid-1980s, three large outbreaks of multi-drug-resistant TB have been investigated by the CDC, all associated with HIV patients and all involving transmission to staff.[33]

Thousands of health care workers around the country will be exposed to TB occupationally, and thousands will see their skin tests converted to positive. While a small percentage will develop the active disease immediately, a positive converter will face a 10% risk if not treated and an almost a zero risk if treated; however, if the strain is multidrug resistant, the potential for developing active disease again increases. These positive-conversion health care workers will be put on an antibiotic treatment for 6 months. In some cases, hepatitis is a side effect of the treatment. The full extent nationally of health care worker conversion to TB and the number of cases of active disease are unclear,[28] due to unreliable data and unreliable reporting.

COLLECTIVE BARGAINING

Regulatory agencies provide minimal protection for health care workers, as proved by the lack of specific standards and regulations. The labor unions responsible for the health care worker sectors should negotiate, within the collective agreement infrastructure, stronger health and safety agreements that could potentially make up the vacuum left by the lack of regulation. Though health care worker unions have made major health and safety inroads and continue to do so, collective agreements do not reflect language that legally binds health care employers in health and safety.

A review of a sample of collective agreements (Service Employees International Union [SEIU] Local 790 and City and County of San Francisco, July 1992–April 1994; SEIU and Local 790, and City and County, May 1, 1994 and April 30, 1996; Staff Nurse and Per Diem Nurse, SEIU, and City and County Locals 250 and 535, 1994 to 1996) negotiated by major health care trade unions and their locals reveals an average of one or two pages of language but nothing dealing with any of the "hot-button" risk issues that affect health care workers. The collective agreements that were scanned did cover "right to refuse dangerous work" or the creation of health and safety committees, but little else relating to health and safety was mentioned.

The majority of health care workers are still unorganized. Unionization in the health care industry declined during the 1980s. The percentage of health care workers who belong to unions dropped from 15% of the work force in 1984 to 11% in 1992. If current trends continue, it is projected that by the year 2000, less than 10% of health care workers will be organized. Rank-and-file participation in health and safety on a national and local level is still the dream, not the reality. There have been no strikes solely for health and safety. Health care union health and safety departments are understaffed, and at least one major health care union is downsizing its health and safety staff.

Health care is undergoing major shifts. In 1970, two thirds of all health care workers were employed by hospitals. Today that percentage has dropped to 50%. This work force remains largely unorganized.

LACK OF TRAINING APPROPRIATE TO RISK

Health care workers are poorly prepared to undertake the hazardous duties for which they are being hired. Training on the range of occupational risks does not come close to the level of risk. For example, nurses receive an average of 4 h of training in nursing school on the biomechanics of lifting and transferring patients. According to a study in the *Home Health Care Services Quarterly*, home health care aides have an annual low-back injury rate of 15.4 per 100 FTEs, compared with a rate of 5.9 for hospital nurses' aides.[34] Training for frontline nurses in transferring patients does not reduce the risk, according to many studies.[35-39]

Nursing schools do not offer any occupational health prevention courses in their curricula, and no specific standards for health and safety for student nurses exist. Only 4.2% of nursing schools offered their students the hepatitis B vaccine.[40] Nursing students are not covered by OSHA. Medical school curricula do not cover—in any specific way—the occupational risks faced by doctors in the clinical setting.

CONCLUSIONS

Health care worker injury has risen to epidemic levels. Webster's defines epidemic as "affecting many individuals within a population." The data presented in this chapter support the terminology of epidemic. Ergonomic injury, needlesticks, infectious diseases, chemical exposures, and introduction of new medical technology and chemical treatments without testing for occupational risk all contribute to the upward trend in health care worker injury and illness. Underreporting rates of injury in health care, probably as high as in other industries that can run as much as 50%, puts these data in an even darker light. Looking at the data as a whole instead of in separate categories shows an industry-wide injury and risk problem. Ironically, little knowledge of this epidemic exists within the health care community.

Approximately 1.5 million health care workers will be exposed to a needlestick on the job. There is a low prevalency rate of seroconversion to HIV (rate of 0.42%),[41] but the stress that health care workers suffer as a result of needlesticks has not been qualified. The anxiety of seroconversion to HIV or hepatitis has been known to affect health care workers deeply,[42] though the majority still accept risks as part of the job.

The solutions to this industry-wide problem are not simple. They would require more legislation; regulation; unionization of health care workers; collective bargaining language reflecting the occupational risk; more inspections by regulatory agencies; increase in in-house safety personnel in health care facilities; labor–management health and safety committees; and commitment from the safety design industry to produce "passive" safety devices for health care workers. Standards specifically for health care need to be written, and greater funding must be allocated for employee health and safety departments. All these solutions must be done in a growing climate of "corporate structuring of medicine," where profit and reductions outweigh safety.

ACKNOWLEDGMENT

This article originally appeared in *New Solutions,* 1997; 7(3): 81–88 by William Charney, past Director of Environmental Health, San Francisco Hospital, San Francisco, CA.

REFERENCES

1. Makofski D: *Epidemiological Analysis of Occupational Injury in Health Care. Essentials of Modern Hospital Safety.* Vol. 3. New York, CRC/Lewis Press, 1995; 18.
2. U.S. Department of Labor Bureau of Labor Statistics: Worker Safety Problems Spotlighted in Health Care Industry. Publication 94–6. 1994.
3. Sterling D: *Overview of Health and Safety in Health Care. Essentials of Modern Hospital Safety.* Vol. 3. New York, CRC/Lewis Press, 1995; 1.
4. Charney W: Lifting Team: A Design Method to Reduce Lost-time Injury in Nursing. *Am J Occup Health Nursing.* 1991; 39(5): 231–234.
5. Jensen R: Low Back Pain in Nursing. *J Saf Res.* 1983; 19: 21–25.
6. Stubbs D: Back Out: Nurse Associated with Back Pain. *Int J Nursing Stud.* 1986; 23: 235–236.
7. Grzydowski M: Prevalence of Anti-Latex IGE Antibodies Among RNs. *J Allerg Clin Immunol.* 1996; 98(3): 535–544.
8. U.S. Department of Labor Bureau of Labor Statistics: Workplace Injuries and Illnesses in 1994. Publication 95–508. 1994.
9. U.S. Department of Labor Bureau of Labor Statistics: Survey of Occupational Injuries. Publication 96–11. 1996.
10. U.S. Department of Labor Bureau of Labor Statistics: Publication 96–163. 1996.
11. Owen B: *An Ergonomic Approach to Reducing Back Stress in Nursing Personnel. Essentials of Modern Hospital Safety.* New York, CRC/Lewis Press, 1995; 333.
12. Service Employees International Union: *Caring 'Til It Hurts.* Washington, D.C. 1995.
13. Alter M: Hepatitis in HCW. *J Infect Dis.* 1982; 145: 806.
14. U.S. Department of Labor Bureau of Labor Statistics: *Fatal Workplace Injuries in '94.* Publication 908. 1996.

15. Meyer A: Low Back Injuries, Home Health Care. *Home Health Care Serv. Q.* 1993; 14(2/3): 149–155.

16. U.S. Department of Labor Bureau of Labor Statistics, OSHA Report. Data available upon request. Washington, D.C.

17. Joint Commission on Hospital Accreditation Organization, Background on Accreditation Process and Data. Cited by M. Millenson in *Chicago Tribune.* December 3, 1989; 3.

18. Perry M: Chemotherapy Toxicity and the Clinician. *Semin Oncol.* 1984; 9: 1–4.

19. Finucane E: Operating Room Safety. *Essentials Mod Hosp Saf.* 1990; 1: 99.

20. Bellows J: Ribavirin Aerosol. *Essentials Mod Hosp Saf.* 1990; 1: 99.

21. NIOSH Health Evaluation Report. HETA. Publication 85–126. 1985.

22. Hendrick D: Asthma in Hospital Staff. *Br Med J.* 1975; March 15: 607–608.

23. NIOSH Health Evaluation Report. HETA. Publication 83, 074–1525. 1983.

24. Centers for Disease Control and Prevention (CDC): Evaluation of safety devices for preventing percutaneous injuries among health care workers during phlebotomy procedure. 1997; 46(2): 21–25.

25. Klein R: Hepatitis C in Dentists. *Lancet.* 1991; 330: 1539.

26. Lanpthen B: Hepatitis C in HCW. *J Infect Control Hosp Epidemiol.* 1996; 9(12): 548–552.

27. Centers for Disease Control and Prevention (CDC). What is the Risk of Hepatitis C for HCW. Hepatitis Surveillance Report No. 56. 1996.

28. Bowden K, McDiarmid M: Occupationally Acquired TB. *J Occup Med.* 1994; 36(3): 320–325.

29. Catanzano A: Nosoconical Transmission of MDR in New York and Florida 1988-1991. *MMWR.* 1991; 40(34): 585–591.

30. Brema C: TB in Chronic Care Facility. *J Infect Control Hosp Epidemiol.* 1995; 9(12): 548–552.

31. Centers for Disease Control and Prevention (CDC). Nosocomial Transmission of MDR in New York and Florida 1988-1991. *MMWR.* 1991; 40(34): 585–591.

32. Centers for Disease Control and Prevention (CDC). MTB Transmission in a Health Clinic, Florida 1988. *MMWR.* 1989; 38: 256–258, 263–264.

33. Centers for Disease Control and Prevention (CDC). Nosocomial Transmission of MDR TB, New York and Florida 1988–1991. *MMWR.* 1989; 38: 256–263.

34. Meyers N: Lower Back Injuries Among Home Health Care Aides. *Home Health Care Serv Q.* 1993; 14(2/3): 149–155.

35. Veaney P: Back Injury Prevention Among Personnel, Role of Education. *Am J Occup Health Nurses.* 1988; 36: 327–333.

36. Stubbs D: Back Pain in Nursing Profession. *Ergonomics.* 1987; 26: 767–779.

37. Wood D: Design and Evaluation of Back Injury Prevention Programs in a Geriatric Hospital. *J Spine.* 1987; 12: 77–82.

38. Snook S: Study of Back Prevention Approaches to Low Back Injury. *J Occup Med.* 1978; 20: 478–481.

39. Strather C: Work and Back problems. *J Occup Med.* 1990; 409: 75–79.

40. Pearl J: *Nursing Student H&S Training. Essentials of Modern Hospital Safety.* Vol. 3. New York, CRC/Lewis Press, 1995; 460.

41. Rythorne M: Surveillance of HCW Exposed to Blood from Patient Infected with HIV. *N Engl J Med.* 1988; 319: 1118–1123.

42. Husak PJ: Psychological Effects on Health Care Workers After Needlesticks. To be published.

2 Health Hazards in Nursing and Health Care: An Overview

Bonnie Rogers

CONTENTS

INTRODUCTION

It has become apparent that health care workers are exposed to a variety of occupational insults that can result in serious acute and long-term adverse health outcomes.[1-10] The risk of occupational illness and injury in this occupational group is likely to rise as demands for health care increase, with concomitant increases in length of working hours, shift rotation, and increased workload of patients requiring highly complex technological care and services. This chapter will present an overview of significant work-related hazards faced by health care workers including exposure risk factors. The importance of prevention and control strategies will be emphasized.

According to the U.S. Bureau of Labor Statistics (BLS), the health care industry is one of the largest employers in the United States having more than 7 million workers in 1990, with that number expected to increase to nearly 11 million by the year 2000.[11] While hospitals employ nearly 50% of all health service sector workers, the remainder work in community settings including offices, laboratories, and outpatient facilities, which employ almost 30%, and in nursing and personal care facilities, which employ 20%.[12] Approximately 75–80% of those employed are women, many of whom are in the childbearing age group. Health care environments pose a wide range of hazardous exposures, thereby placing workers in health risk situations.[1,13]

The U.S. Department of Labor[2] reports that nurses' aides and orderlies are the largest employee group among health care workers, followed by registered nurses (RNs) and then licensed practical nurses (LPNs). RNs are the largest professional health care employee group with more than 2 million workers. The U.S. BLS[2] reports the incidence rate per 100 full-time workers for nonfatal occupational injuries and illnesses for 1993 was 11.8 for hospital establishments and 17.3 for nursing and personal care facilities. This compares to a private industry rate of 8.5. In addition, the BLS reported that for 1992, the incidence of injuries and illnesses involving days away from work was greater for certain occupations such as nursing care and housekeeping services. Nurses' aides reported 111,100 cases of disabling injury and illness, second only to truck drivers who reported 145,900 cases. For persons in all occupations working less than 1 year, nurses' aides were reported as having the most injuries/illnesses (approximately 67,000), primarily sprains and strains, and overexertion related to patient care.[2]

While studies have been conducted in targeted hospitals for selected hospital hazards, the perceived types and degrees of hazards and risks and their management are not well understood. In addition, data concerning occupational risk hazard assessment in community health care settings (e.g., home health agencies and private practice offices) are virtually absent. However, evidence does exist that nurses working in private oncology offices who handle antineoplastic agents are at greater risk of exhibiting urine mutagenicity, symptomatic complaints, and reproductive toxicity than are hospital-based nurses.[1,8,14]

Healthy People 2000[15] recognizes an increase of injuries among nursing personnel and the importance of preventing occupational health hazards in terms of reducing both short-term and long-term illness and injuries. Keeping the workforce safe and healthy is important to the health and safety of our nation and is mandated by the Occupational Safety and Health Act of 1970.

OCCUPATIONAL HEALTH HAZARDS

As a result of specific agent exposure, occupational hazards are classified in five categories, including

1. Biological/infectious hazards—agents, such as bacteria, viruses, fungi, or parasites, that may be transmitted via contact with infected patients or contaminated body secretions/fluids
2. Chemical hazards—various forms of chemicals that are potentially toxic or irritating to the body system, including medications, solutions, and gases
3. Environmental/mechanical hazards—factors encountered in the work environment that cause or potentiate accidents, injuries, strain, or discomfort (e.g., poor equipment or lifting devices, and slippery floors)

4. Physical hazards—agents within the work environment, such as radiation, electricity, extreme temperatures, and noise that can cause tissue trauma
5. Psychosocial hazards—factors and situations encountered or associated with one's job or work environment that create or potentiate stress, emotional strain, or interpersonal problems

BIOLOGICAL/INFECTIOUS AGENTS

Exposure to biological agents and subsequent infections that develop are the most familiar occupational risk faced by health care workers. Infections from viral organisms present the greatest risk; however, some are now vaccine preventable. Of greatest concern are the infectious agents, human immunodeficiency virus (HIV), hepatitis viruses, rubeola (measles), rubella (German measles), herpes viruses (herpes simplex), varicella (chicken pox/shingles), and cytomegalovirus (CMV), and the bacterial agent *Mycobacterium tuberculosis* (TB).[128]

Risk of exposure to biological/infectious agents heightened in 1984 when a new era of concern about occupational hazards for health care workers began with the first report of HIV, transmitted by needlestick injury.[16] Exposure risk increased the awareness in the health care community that numerous biological agents may be transmitted via needlestick and mucous membrane exposure, and that work-related exposure to blood-borne pathogens, including HIV, hepatitis B virus (HBV), hepatitis C virus (HCV), and other potentially infectious agents are a significant threat.[17–19, 132]

While occupational exposure may occur primarily percutaneously from needlestick or sharps injury or from mucous membranes contact with contaminated blood/body fluid, percutaneous exposure is the principal route accounting for 86% of the occupational HIV exposure.[20,133] According to the Centers for Disease Control and Prevention (CDC), 47 health care workers in the United States have been documented as having HIV seroconversion following occupational exposure. Of these, 42 had percutaneous exposure, 5 had mucous membrane exposure, 1 had both percutaneous and mucous membrane exposure and 1 had an unknown exposure route.

Numerous studies have found that nurses are a primary target of needlestick injuries,[20–25] and several studies have reported incidence rates for needlestick injuries ranging from 10–34%.[26–28] However, these figures may be seriously underestimated because studies also indicate estimates ranging from 30–60% of failure by nurses to report needlesticks.[26,29–31] Marcus and co-workers[20] report that needlestick exposure to HBV carries a 6–30% risk of infection while HIV needlestick contact carries less than a 1% chance of seroconversion. The CDC estimates that between 15 and 30% of health care workers with frequent blood contact have one or more serological markers of HBV infection and that 1–2% of these persons are chronic carriers of HBV.[17]

In contrast, health care workers with no blood contact have HBV marker and carrier rates similar to those of the general population. However, in a review of the literature on HBV, HIV, and HCV exposure, Gerberding[32] reports that the risk of transmission of HBV and HIV after needlestick injury seems to be related to the level of viral titer in the contaminant, and for HBV, correlates with the presence or absence of hepatitis Be antigen (HBeAg). The author reports on estimates of HBV infectivity ranging from 2% (HBeAg absent) to 40% (HBeAg present) and for HIV from 0.2–0.5%. The outcome of HBV exposure is described by the CDC (1996), which has estimated annual infections for 1984–1994 to range from 140,000 (1994) to 320,000 (1984). In 1994, the CDC estimated the number of HBV infections in health care workers to be 1000, with an alarming 150 deaths per year.[33,44,132]

Much of the concern related to needlestick injuries is related to the possible consequences of contracting HIV and ultimately acquired immunodeficiency syndrome (AIDS). As of December 1995, the CDC reported that in the United States alone, 513,486 persons had been diagnosed with AIDS (nearly 99% of these reported in adults and adolescents).[133] The CDC reports that this figure may be lower than the actual amount because of inadequate diagnoses and nondiagnoses due to individual fear of positive test results. As of December 1993, the CDC identified 123 documented or possible cases of occupationally acquired HIV, which represented two thirds of the 176 documented cases worldwide.[34] Nurses and clinical laboratory workers, primarily phlebotomists, ranked first among HIV infected workers, with each group accounting for 24% of the 123 cases.[19] As of December 1995, the number of documented or possible work-related HIV infected persons was reported at 151.[133]

Work practice behaviors may lend themselves to increased risk of injury or, if effectively practiced, risk reduction. Approximately 800,000 needlestick injuries occur to health care workers annually in the United States, and about 16,000 have a potential risk for HIV transmission.[148] While the rate of HIV seroconversion is only about 1 in 300 from an HIV-contaminated needle, the HBV infection rate from an HBV-contaminated needle is about 1 in 6.[148] Results of a national survey of certified nurse midwives indicated that 36% of respondents reported breaking needles and 63% (453/721) continued to cap needles,[28] even though evidence is clear that recapping is the cause of one third of all needlestick injuries.[35] Another study of four Michigan hospitals found that up to 60% of needles were found to be capped on inspection of sharps disposal containers.[36]

English[38] reported findings of a study on needlestick injuries conducted for a 1-month period in 17 hospitals in Washington, D.C. Of the 72 injuries reported, 46% occurred in RNs and recapping was associated with most of the injuries (14%). Eighteen injuries (25%) were to "downstream" housekeepers and aides who did not use such devices in their practice.[39] Mallon reported similar findings in a study of 332 reports of occupational blood and body fluid exposure during a 9-month period where needlestick and other sharps injuries accounted for 83.4% of all reports. In addition, failure to use universal precautions was cited in 34% of the reports. Jagger[22] reported surveillance data on percutaneous injuries for a 1-year period (1992–1993) from 58 participating hospitals. Of all the cases reported (n = 471), nurses and phlebotomists each accounted for 157 and 150 injuries (33.3 and 31.8%) respectively, with two thirds of the injuries occurring in patient rooms (53.5%) and in the emergency department (14.0%).

Recapping needles has been identified as a continual source of worker exposure, and many of these injuries are considered to be preventable through use of safer devices, procedures, and work practices.[41,134,135] Younger and colleagues[136] reported a significant decrease in the rate of needlestick injuries from 14 to 2 per 100,000 inventory units in health care workers who used a (3-cc) shielded syringe. Prince and colleagues[137] reported a 75% reduction in needlestick injuries after introducing a needleless system. Makofsky and Cone[135] also reported significant reductions in the recapping of needles (as measured through counting contaminated-only needles placed in disposal containers) simply by placing disposal containers in a more convenient location near the patient's bed. The effectiveness of work practice controls is of concern, and the use of engineering controls would be more effective.[37]

Exposure to HCV is also of concern. HCV is one of the many viruses that used to be called non-A, non-B hepatitis and is responsible for the majority of bloodborne non-A, non-B hepatitis worldwide, accounting for 25–37% of the reported cases of viral hepatitis.[42,141] While exposure to HCV is mostly reported with parenteral drug use, HCV is occupationally transmitted mostly through large repeated percutaneous exposures to blood as a result of needlestick or sharps injury.[138,140] Those with acute HCV infection will most probably develop chronic liver disease.[139] The CDC estimated the number of HCV infections in 1994 to be 35,000 with the number of reported cases at 4,470. Nearly 4 million persons are estimated to have chronic infection with the annual attributable deaths to chronic liver disease estimated at 8,000–10,000. The number of infections among health care workers in 1994 is not given.[132]

Estimates for HCV transmission are not as well documented, and persons working in emergency rooms, dialysis units, blood banks, and operating rooms are potentially at greatest risk. In one study of patients receiving hemodialysis, HCV antibody prevalence was found in 11 of 90 patients (12%) but in none of the staff.[43] In a conflicting report, Forseter et al.[141] conducted a cross-sectional serum-generated study of patients and hemodialysis nurses to assess the risk of HCV infection for the nurses. While 19% of patients were HCV seropositive (24/125), only 1 of 56 nurses was HCV seropositive. However, in a hospital-based study, for a period of 10 years, Lanphear and co-workers[142] reported findings that the incidence of HCV in exposed health care workers was three times that of the general population. Most exposures occurred through needlestick injury and exposures were twice as likely to occur in the emergency department. While health care worker risk of HCV exposure is somewhat greater than in the general population, the cumulative risk appears to be low relative to that of HBV.[142]

Health care workers exposed to HCV should be tested immediately post-exposure and again 6–9 months later to detect HCV antibodies. At least 90% of those infected will have HCV antibodies by 9 months. If results are positive, chronic hepatitis should be determined and treatment referral should be implemented.[142]

Hepatitis A virus (HAV) is transmitted primarily by the fecal–oral route with an incubation period of 2–6 weeks; there is no chronic carrier state. Transmission from infected patients to health care workers has been reported to occur from patients with diarrhea, primarily as a result of poor hygiene practices on the part of the health care worker.[45,46]

Large hospital outbreaks of rubeola were relatively uncommon during the 1970s and 1980s, but became a major concern by the end of the 1980s when outbreaks were occurring in health care settings, and inadequate immune status was detected in health care workers.[47] Most health care agencies now require measles immunity as a condition of employment, and most schools of nursing have similar requirements for admission. The measles, mumps, rubella (MMR) vaccine is readily available.

Reasons for concern about rubella in health care workers are twofold: not only are workers susceptible for acquiring rubella, but also they can transmit rubella to other susceptible health care workers, patients, or fetuses during gestation. While the number of nosocomial outbreaks of rubella has declined, universal immunization against rubella as recommended by the CDC for all health care workers is essential to control any spread.[48] Rubella immunity is now a condition of employment in most health care agencies.

The herpes viruses represent a family of viruses that are ubiquitous and as concerns occupational risk include most notably herpes simplex virus (HSV), varicella-zoster virus, and CMV. Most adults have been infected with at least one form of the virus. HSV is generally self-limiting in the acute phase but the disease is chronic. The virus is transmitted by direct contact with herpetic lesions, primarily of the mucous membranes. HSV causes herpetic whitlow and other localized infections in health care workers.[49] In addition, health care workers can transmit the virus to patients under certain circumstances, such as skin-to-skin contact that occurs when performing patient-care activities, including turning or bathing patients. The CDC recommends that personnel with herpetic whitlow not have direct contact with patients until lesions are healed, and that personnel with orofacial herpes be restricted from care of newborns, patients with burns, or severely immunocompromised patients.[50,145] The use of gloves and gowns and appropriate attention to hand washing, hygiene, and equipment care must be emphasized. The reader is referred to *Guidelines for Isolation Precautions in Hospitals* for a detailed and comprehensive discussion.[145]

Varicella-zoster virus is the etiologic agent for chicken pox, which is transmitted through respiratory droplets. This is of particular importance to susceptible health care workers, because persons who have not had chicken pox can be infected from persons with chicken pox or shingles.[51] A vaccine for varicella is now available, and barrier precautions, such as gloves and gowns, and appropriate room ventilation for infected patients should be emphasized. Restriction of susceptible health care workers from direct patient care may be needed.

CMV infection is widespread among health care workers with prevalence rates in the United States ranging from 43–79%.[52,143] Sources of exposure include prolonged or multiple contact with infected secretions such as urine, saliva, breast milk, cervical secretions, and respiratory secretions. Some reports of nosocomial CMV infections have indicated that nurses and others caring for patients were not at increased risk for nosocomial CMV transmission, and that community and family sources appeared to be more likely sources.

Persons at increased risk for both high prevalence of infection and CMV shed in secretions are young children, immunocompromised patients (including those with organ transplant, leukemia, burns, or kidney failures), persons with HIV infection, and those actively treated with immunosuppressive drugs. Risks to health care workers caring for these patients center primarily on health care workers who are immunocompromised or are pregnant, or whether health care workers who may be pregnant should be assigned to care for patients known or likely to be excreting CMV (e.g., organ transplant patients and persons with AIDS).[144] The greatest risk is posed to unborn babies of pregnant women who have never had CMV and become infected with the virus. The infant's risk of developing congenital CMV disease can result in hearing loss, visual impairment, mental retardation, and liver disease.[144]

TB is an infectious disease caused by *M. tuberculosis bacteria*. In the late 1980s, the picture of TB in the United States changed dramatically with a significant increase in the rate of new cases, largely related to the epidemic of HIV. Once infected with *M. tuberculosis*, persons with HIV infection often progress quickly to active pulmonary disease, and many are infected with strains of *M. tuberculosis* that are drug resistant.[54]

As reported by the CDC, four hospital outbreaks of multi-drug-resistant noso-comial transmission of TB occurred between 1988 and 1991,[55] with eight cases of TB reported among health care workers in these hospitals. Of these, five health care workers had known or possible exposure to the outbreak cases. Five of the eight health care workers were known to be HIV-antibody positive; one was known to be HIV-antibody negative; and the HIV-antibody status for two was unknown. From two of the hospitals, a tuberculin skin test conversion rate of 37% (19/51) was reported. Several related outbreak characteristics included delayed diagnosis and infection control precautions, poor ventilation, and lack of isolation enforcement. In a later report from one of the outbreak hospitals,[146] control measures recommended by the CDC,[54] including early identification and treatment of active TB, environmental control of infectious particles, and frequent monitoring of health care workers, were instituted and then evaluated 2 years later. Skin test conversions among health care workers dropped from 28% to 0.

The TB news is looking better. During 1995, the CDC reported a 6.4% decline in TB cases from 1994.[147] This represents the third consecutive year (since 1992) the number of TB cases has decreased, resulting in the lowest reported TB rate since surveillance data began in 1953. The CDC attributes this substantial decline to at least six factors:[147]

1. Improved laboratory identification methods of *M. tuberculosis*
2. Broader use of drug susceptibility testing
3. Expanded use of preventive therapy in high-risk groups
4. Decreased transmission of *M. tuberculosis* in congregative settings (e.g., hospitals, and correctional facilities)
5. Improved follow-up of persons with TB
6. Increased federal resources for state and local control efforts

The Occupational Safety and Health Administration (OSHA) has published a draft proposal TB standard, incorporating most of the CDC guidelines. The standard is intended to be published within the year.

CHEMICAL AGENTS

A variety of chemical agents exist in the work environment, particularly in the hospital, although other settings such as nursing homes and private practice offices may be "hotbeds" of exposure to which health care workers are routinely exposed. Whether an individual is at risk of exposure will depend on type, dose, frequency and duration of exposure, work practices, and the individual's health and susceptibility.[2]

Chemical exposure to health care workers can occur through inhalation, ingestion, and absorption through the skin or mucous membranes. Some of the more common hazards used in the health care environment include exposures to disinfecting/sterilizing agents such as glutaraldehyde and ethylene oxide (ETO), chemotherapeutic agents, waste anesthetic gases, and latex.

Chemical disinfectants and cold sterilants such as 2% alkaline glutaraldehyde, which is used to disinfect instruments, can provoke reactive airway symptoms and skin problems, and mercury exposure can result in numerous organ system reactions.[1,56,127] ETO is a chemical agent with mutagenic, carcinogenic, and explosive properties used to sterilize heat-sensitive equipment (e.g., plastics).[57] ETO is a potent sterilizing agent used in operating rooms, anesthetic areas, dental and surgical clinics, and the like, thereby creating numerous settings for exposure.[58] Nearly 100,000 hospital personnel are estimated to have direct or indirect contact with ETO each year. Cancer and reproductive toxicity in animal studies have been reported.[59]

Antineoplastic agents represent a significant hazard for nurses who handle these substances.[10,60–62] Acute health symptoms, such as lightheadedness, nasal sores, nausea, hair loss, facial flushing, depressed leukocytes, and skin rash, have been reported. In a case control study, Selevan and co-workers[63] reported significantly higher fetal loss among nurses occupationally exposed to antineoplastic agents during the first trimester of pregnancy. Christensen and colleagues[64] surveyed 457 female pharmacists concerning work practices related to handling antineoplastic agents. Results indicated that vertical flow hoods were absent for nearly 30% of the respondents and more than 40% of the group reported cytotoxic drug skin contact. Nurses most at risk for toxicological effects are those with regular cumulative exposure to these agents in practice settings such as hospital oncology floors, oncology units, private physicians' offices, and outpatient clinics, the latter two groups having a greater risk of urine mutagenicity.[60] This suggests that duration of exposure and type of agent exposure are significant variables.

Waste anesthetic gases have been studied in numerous epidemiological surveys that suggest these gases are related to various adverse health effects including hepatic and renal disease, and central nervous system and immune system dysfunction.[65,66] Several studies indicate that individuals exposed to waste anesthetic gases have a higher risk of spontaneous abortions and congenital malformations in their offspring. Somatic complaints and leukocytic reduction have also been demonstrated. Exposure to anesthetic gases may result from improperly inflated endotracheal tubes, poorly fitting patient facial masks, and failure to properly connect the anesthetic gas lines or waste disposal lines.[2] Engineering controls such as effective ventilation and scavenging systems and work practices related to proper disposal techniques are effective strategies to reduce exposure.[65]

Latex allergy, which is associated with the use of natural rubber latex gloves, is a growing problem in health care workers, with prevalence rates of 10% reported in heavy glove users, such as operating room personnel.[149] Latex is composed of pure rubber and water with small amounts of resins, proteins, sugar, and mineral.

During the process of making latex products other chemical products are added; however, the protein content is the major sensitizing antigen.[150] Exposure can occur through direct skin contact or inhalation of the allergen. Latex protein binds to the glove powder during the manufacturing process and may be expelled into the air when gloves are donned or removed and subsequently inhaled.[152] Reactions can range from local contact dermatitis to systemic reactions and anaphylaxis.[151,152]

Health care workers at risk for latex exposure should have a complete assessment and be referred for medical evaluation and skin testing if indicated.[153] Workers with latex sensitivity should be provided with non-latex gloves and job transfer, if necessary.[152]

ENVIRONMENTAL/MECHANICAL AGENTS

Environmental and mechanical agents are associated with injuries and accidents in the work setting. Poorly designed or inadequate equipment or work stations can cause discomfort and potentiate injuries, such as cumulative trauma disorders, and cluttered and slippery floors (from spilled fluids) contribute to falls and other accidents. Back injuries are the most expensive workers' compensation problem today. The extent of low-back pain and injury in nursing personnel is thought to be underestimated.[67] A survey by Owen of 503 nurses found that only 34% of respondents with work-related low back pain filed an injury report, and 12% were contemplating leaving the profession because of the problem.[68]

Several studies indicate that tertiary care hospital staff have reported work-related back pain and injury and implicate lifting techniques, poor staffing, ergonomics, lack of use of assistive devices, inadequate communication, and constitutional factors as contributory factors.[69-74] Hospital and long-term care nursing staff are at high risk for sustaining back injuries because they must lift, bend, and pull while moving and transferring patients. Owen and Garg[75] evaluated 38 nursing assistants with respect to factors associated with back injury. Tasks were categorized into 16 areas and activities associated with transferring clients from one location to another (e.g., transfer from toilet to chair; bed to chair; and tub to chair) were ranked highest (top 6 out of 16 categories). Attention to worker–job capabilities was cited as essential.

Wilkinson and colleagues[13] retrospectively investigated occupational injuries among 9668 university health science center and hospital employees during a 32-month period. During this time, 1513 injuries were reported with the most frequent injuries being needlestick injuries (32.1%) followed by sprains and strains (17.2%); of the latter, 55% involved back injuries that were reportedly caused by lifting and twisting motions. Nearly 10% of those injured lost time from work. Out of 18 job classifications analyzed, nursing, which included nurses' aides, ranked third highest in injury reports. In overall reporting, professional nurses reported the highest number of illnesses and injuries; however, within the total nursing group, nurses' aides reported the highest injury attack rate when adjusted for size of population at risk. Workers' compensation costs were highest for back injuries and equaled $171,957. Other types of injuries reported included lacerations and contusions.

Kaiser-Permanente Medical Centers in Portland, OR found back injury rates from workers' compensation claims ranging from 10–30% on hospital units and as a result instituted a back injury prevention project.[76] Divided into intervention and control groups, 55 nurses, nurses' aides, and orderlies participated in the project. Instructional methods on body mechanics, transfer maneuvers, exercise, and stretching were provided to the intervention group. Prior to the intervention, data indicated that 86% reported work-related back fatigue; 74% reported that back pain interfered with the quality of work performance; 55% reported lost work time; 32% reported lost time of more than 3 days related to the injury; 60% required medical intervention; and 91% reported that patient handling put them at risk. "New" nurses with less than 5 years of experience reported more frequent back pain, but veteran nurses considered the injury more disabling. Body mass index and lack of flexibility were significantly associated with back pain and injury. Although not statistically significant, scores to measure back pain and fatigue were reduced post-intervention but no reduction was seen in the control group.

One study has linked nurse–patient ratios to the incidence of back pain or injury.[77] Nurses working on general hospital (GH) units and oncology department (OD) units were compared for musculoskeletal disorders. The nurse–patient ratio for GH nurses was 0.57 for every one patient compared to 1.27 for OD unit nurses. For nurses working on the GH unit, 48% reported work-related back pain and 19% had lost work time compared to 33 and 9% of the OD nurses, respectively. Both groups of nurses were subject to frequent and heavy lifting, lowering, pushing–pulling, and so on.

Greenwood[78] completed a retrospective study of 4000 back injury reports of hospital employees including dietary, housekeeping, RNs and LPNs, and nurses' aides. Of the back injury cases, 40% were represented by nurses' aides. Influencing factors included employment for less than 1 year and long working hours. Nearly 40% of the cases resulted in lost work time. In a related study, Venning and co-workers[79] surveyed and observed registered nurses, nurses' aides, and orderlies and found several factors as significant predictors: jobs with patient lifting, frequency of lifting, job category of nursing aides and orderlies (were nearly 2 times more likely to sustain an injury), and history of previous back injury.

In a British study of work-related back pain, Newman and Callaghan[80] analyzed data from 173 nurses and midwifery staff (65% response rate) with 76% having experienced pain in the last 2 years. The total number of days nurses were "unfit" to work totaled 2,769 or 2.63 days per nurse per year with an estimated cost for lost work time equaling $146,622. Only 6% of nurses indicated they reported every back pain injury they suffered. Neuberger and co-workers'[81] analysis of workers' compensation claims for a 1-year period indicated that 855 injury events were reported, resulting in 4,825 lost work days and $326,886 in medical care and disability claims costs. The overall annual reporting rate was nearly 20/100 full-time equivalent (FTE) workers and occurred more often in women employees under 30 years of age. In fact, the rate in younger employees was 77% higher than in older employees. Repeat injurers had nearly twice as many lost work days and accounted for nearly one-fifth of those injured, one-third of the reported incidents, and one-half of the claims costs.

VIOLENCE

Violent behavior toward medical and nursing personnel by hospitalized medical, surgical, and psychiatric patients and by persons seeking care in emergency departments has been well documented.[92–94,126,129] Still, it is considered underreported. Victims of nonfatal violence are increasingly caregivers in nursing homes and hospitals who are injured by patients.[95]

Carmel and Hunter[195] conducted a 1-year study of injuries in a 973-bed California maximum security hospital and reported staff injuries at a rate of 16 per 100 staff compared to 8.3 per 100 general full-time workers. Jones[96] reported that in a Veterans Administration (VA) medical center, nursing assistants, followed by nurses, and physicians were most likely to be assaulted. Lavoie and co-workers[94] reported on results of a survey of 127 emergency room medical directors. Of the sample, 43% indicated at least one physical attack on a medical staff person per month, which included violent acts that resulted in death. Over a 9-year period nearly 5,000 weapons were confiscated from 21,456 patients at a Los Angeles trauma center.

Much of the study of violence in health care settings has focused on psychiatric settings.[97,98] In a study of 154 psychiatric nurses, 74% indicated that staff members could expect to be assaulted by patients during their careers, and in fact 73% had already been assaulted.[99] McCullough and colleagues[100] report that weapon carrying in psychiatric facilities and emergency rooms is not uncommon, and staff members are usually unable to predict who the weapons carriers are.

While the actual costs of violence toward health care workers are not well documented, several investigators have reported lost work time of nearly 50% for those injured.[101] In addition, posttraumatic stress symptoms have been reported by victims of patient assault.[102]

One must ask what factors perpetuate the problem, is the risk of these types of injuries perceived as real, what other types of injuries occur that result in adverse health effects, and what are the occupational hazards in community health-related settings? These issues in all types of settings need careful examination.

PHYSICAL AGENTS

Physical agents in the work environment, such as noise or vibration, can create health hazards through the transfer of physical or mechanical energy to humans in that environment, thereby producing tissue trauma. For example, human exposure to electrical hazards can result in internal and external burns, and gaseous embolism.[2] Exposure to physical agents in the workplace can result in serious consequences.

Radiation is a common physical agent that is widely used in medicine for various diagnostic and therapeutic procedures and protocols. Health care workers may be at risk for radiation exposure during diagnostic X-rays and radioactive implants, and from body fluids of patients receiving metabolized therapeutic nuclear radiation.[103] Radiation exposure has been linked to cancer development and reproductive toxicity.

Light amplification by stimulated emission of radiation (laser) emits nonionizing radiation in the ultraviolet, infrared, or visible spectrum and is effective by pointing a very bright narrow beam of light at the point of impact. Health care workers may be at risk of eye or skin injury if the laser is not calibrated, maintained, or handled correctly.[104]

Noise is another physical hazard that can result in hearing loss to health care workers if decibel levels are high enough.[105] Most health care environments do not have noise exposures above the allowable levels; however, there are some specialized nursing tasks or nursing care environments where noise may be excessive for short periods, or may cause fatigue, nervous tension, and decreased productivity. For example, nurses working in air rescue operations, such as Life Flight, are exposed to considerable noise emitted by a variety of equipment.[106] Exposure to noise can result in hearing loss or psychological stress.[105] The bombardment of noises from equipment, alarms, visitors, suction equipment, ventilators, patients, and call bells, as well as from other health care providers, can lead to sensory overload. Loud noise reported in critical care environments has been associated with loss of concentration, irritability, and physical health problems such as headaches and high blood pressure.[107]

PSYCHOSOCIAL AGENTS

Psychosocial agents or stressors are well known to nursing. They may result in significant amounts of stress leading to a variety of work-related problems such as absenteeism, staff conflict, staff turnover, decreased morale, and decreased practice effectiveness.[108-110] Sources of stress identified in the work environment include poor working conditions, work overload, role ambiguity, organizational politics, physical danger, poor communications, and shift work.[111]

Several studies have been performed on nurses in various nursing specialties to identify stress existence and stress factors. One survey study of 138 intensive care unit (ICU), hospice, and medical–surgical nurses indicated different factors as stressful.[112] ICU and hospice nurses perceived significantly more stress than medical–surgical nurses related to death and dying; ICU and medical–surgical nurses perceived significantly more stress than hospice nurses related to floating; and medical–surgical nurses perceived significantly more stress than ICU and hospice nurses related to work overload/staffing.

In a related study of 561 ICU and non-ICU nurses from 36 units in 16 hospitals, non-ICU nurses had more work pressure, absenteeism, and health complaints than did the ICU nurses.[113] Snape and Cavanagh[114] indicated that major stressful events reported by neurosurgical nurses included being exposed to life and death situations among young children, being short of essential resources, being on duty with too few staff, and dealing with aggressive relatives. These findings were echoed in separate studies of dialysis nurses and pediatric nurses who indicated that work load was a major contributing factor not only to overall stress and work performance, but also to burnout.[115,116] Pediatric nurses also reported that the relapse or sudden death of a favorite patient was their greatest source of stress.

In a survey of 69 emergency nurses, nurses repeatedly cited as particularly stressful inadequate staffing and resources, too many nonnursing tasks, changing trends in emergency department use, continual confrontation with patients and families who exhibited crisis behavior, and shortage of nursing staff (especially during busy periods and at night).[111] A survey of nurses caring for hemophilia patients indicated that failure of patients to take steps to prevent transmission to HIV, fear of getting infected, and repeated loss experienced as patients died from infection, were the nurses' greatest sources of distress.[117] Oncology nurses reported significant amounts of stress associated with their practice particularly related to the terminality of the disease process. Absenteeism, high staff turnover, poor quality control of work, poor industrial relations, and emotional exhaustion were likely to occur as a result of these factors.[119,124]

In addition to stress, several studies conducted have reported that burnout is a significant problem. Contributing factors included shift work, constantly dealing with death and dying, emotional exhaustion, role ambiguity, and workload pressure.[118,120–125,130]

Job stress has been studied more extensively than any other work-related hazard. In health care work environments stress continues to escalate and job dissatisfaction and burnout ensue. In addition, questions arise as to whether stress and injury are interrelated. Work-related events that could predispose to stress and other adverse health-related outcomes should be carefully examined to more clearly delineate appropriate settings and controls to manage these phenomena.

SHIFT WORK

It is estimated that 11.5 million Americans (27% men and 16% women) work some type of shift work pattern.[82,83] The demands of shift work cause desynchronization of the internal rhythms, which can lead to a variety of psychological and physical problems including gastrointestinal disturbances, exhaustion, depression, anxiety, interpersonal relationship difficulties, and higher rates of accidents.[84,85]

Research studies on shift work have reported adverse effects on workers' health, performance, and mental and physical fitness.[86,87] Gold and colleagues[86] conducted a hospital-based survey on shift work, sleep, and accidents among 635 Massachusetts nurses. In comparison to nurses who worked only day or evening shifts, rotators (all three shifts) had more sleep and wake cycle disruption and nodded off more at work. Rotators had twice the odds of nodding off while driving to or from work, and twice the odds of a reported accident or error related to sleepiness.

Shift workers report a lower sense of well-being with lower participation rates in social organizations, engage in more solitary activities, report higher incidences of family and sexual problems, have higher rates of divorce than day workers, and a decrease in work performance. Factors that affect adjustments to shift work include the type of shift work schedule; the frequency of the schedule changes; the degree to which workers adjust to their social, dietary, and sleeping habits to coincide with their shift work schedule; and age of the worker.[88]

La Dou[88] and Scott[89] report that 20% of workers are unable to tolerate night work, and Akerstedt reports that health problems in shift workers usually increases with age. Tepas[90] reports that night shift workers do not get an adequate amount of sleep, leading to fatigue that results in decreased alertness and ultimately decreased quality of patient care. These investigators indicate that rotating shifts work better if they change clockwise, from day, to evening, to night, instead of day, to night, then evening shift. Application of circadian principles to design of hospital work schedules may result in improved health and safety for nurses and patients.[91]

PREVENTION AND CONTROL

Measures can be taken to reduce the risks of exposure to work-related hazards in health care environments. Methods for hazard abatement include

1. Engineering and work practice controls
2. Administrative controls
3. Personal protective equipment

Engineering control strategies can govern worker exposures by modifying the hazardous source or reducing the quantity of contaminants in the worker's environment. Examples of engineering controls include elimination of the hazard, substitution of a less harmful material for one that is hazardous, alteration of a work process to minimize worker exposure and contact, and provision of better and safer designs for devices.[104,131] For example, a safer needle or a needle-free intravascular access device will reduce the likelihood of transmission of bloodborne pathogens such as HBV or HIV. Placing needle disposal containers as near as possible to the point-of-use so that needles do not need to be recapped, and providing waste disposal, hand washing, and laundry facilities are all examples of engineering controls.

In providing nursing care to patients with *M. tuberculosis*, specialized rooms with high-efficiency ventilation filters and ultraviolet lights may be used. Engineering controls for chemical agents focus on substitution of a less toxic chemical agent or reduction of the amount of toxic material. Ventilation is a major engineering control for reducing risks from chemical agents. Engineering controls to reduce worker exposure to physical agents (such as ionizing and nonionizing radiation) include such practices as designating areas and rooms where radiation exposures are likely to occur, and reducing the scatter of rays by the use of shields.

Work practice controls are designed to enhance effective and appropriate work practices, and to alter those practices that place workers at risk. This involves good hygiene and good housekeeping, including waste disposal, and use of assistive devices to alter work practices, such as in situations requiring excessive torque or lift.

Administrative controls are methods for controlling worker exposures by job rotation, work assignment, time periods away from the hazard, and training. When exposures cannot be reduced to acceptable limits through engineering and work practice controls, then efforts should be made to limit the worker's exposure through administrative controls. Administrative controls include monitoring of the workplace and workers to limit exposures, implementation of policies and procedures to reduce occupational exposure risks, routine air sampling for chemical exposures, and proper machine and equipment maintenance.[104,106,131]

Programs to vaccinate workers against vaccine-preventable diseases such as measles, rubella, hepatitis B, influenza, and tetanus; health maintenance, screening, and surveillance programs; and the provision of training and education in risk reduction strategies, such as handling sharp devices and clean up of wastes, can be effective administrative strategies.

The establishment and availability of procedures related to occupational health and safety measures and practices are essential components of effective administrative control. Administrative controls to reduce chemical risks include limiting the amount of exposure time through staffing assignments, rotating shifts to minimize exposure of any individual to excessive exposures, and dividing work projects into smaller increments over time and thereby decreasing exposure time.

Examples of administrative controls to reduce psychosocial stressors include providing programs about death and dying, improved communication systems, and burnout; providing a forum for interaction and communication among staff; and establishing employee assistance programs.[131]

Personal protective equipment (PPE) is the last resort for worker protection and should only be relied on when hazardous conditions cannot be eliminated through engineering, work practice, or administrative controls. PPE used in the health care environment may include gloves (made of latex, vinyl, or other materials), goggles or other eye protection, face masks of various types, gowns, aprons, laboratory coats, and head and foot coverings. Lead shields for personnel working with ionizing radiation equipment, hearing protection for first responders and those working in aircraft transport services, and laser goggles or glasses to protect workers from laser burns are other examples.

There are numerous guidelines and standards established by the OSHA designed to promote and protect worker health; however, there are also personal and administrative strategies that can be effectively utilized. It is essential that all health care professionals take an active role in worker and workplace health promotion to make the workplace a safe and healthful working environment.

REFERENCES

1. Frazier, L.M., Thomann, W.R., and Jackson, G.W. 1995. Occupational hazards in the hospital, doctor's office, and other health care facilities. *North Carolina Medical Journal* 56(5): 1–7.
2. Rogers, B., and Travers, P. 1991. Occupational hazards of critical care nursing: overview of work-related hazards in nursing: Health and safety issues. *Heart and Lung* 20(5): 486–499.
3. U.S. Department of Labor, Bureau of Labor Statistics. 1994. *Issues in Labor Statistics*. Summary 94–10. August 1994.
4. American Association of Critical Care Nurses. 1989. *Handbook on Occupational Hazards*. Newport Beach, CA.
5. Nightingale, F. 1858. Sanitary condition of hospitals. Paper read at the National Association for the Promotion of Social Science, Liverpool, October (Reprinted in *Notes on Hospitals*, 3rd ed.).
6. Patterson, W.B., Craven, D.E., Schwartz, D.A., Hardell, E., Kasmer, J., and Noble, J. 1985. Occupational hazards to hospital personnel. *Annals of Internal Medicine* 102: 658–680.

7. Valenti, W.M. 1986. Infection control and the pregnant health care worker. *American Journal of Infection Control* 14: 20–27.
8. Valanis, B., and Shortridge, L. 1987a. Self protective practices of nurses handling antineoplastic drugs. *Oncology Nursing Forum* 14: 23–27.
9. Votra, E.M., Rutala, W.A., and Sarubbi, F.A. 1983. Recommendations for pregnant employee interaction with patients having communicable infectious diseases. *American Journal of Infection Control* 11: 10–19.
10. Rogers, B. 1987b. Work practices of nurses who handle antineoplastic agents. *American Association of Occupational Health Nurses Journal* 35: 24–31.
11. U.S. Department of Labor, Bureau of Labor Statistics. 1991. *Occupational Injuries and Illnesses in the United States by Industry*, 1989. Bulletin 2379.
12. U.S. Department of Labor, Bureau of Labor Statistics. 1994. *Issues in Labor Statistics*. Summary 94–6. June 1994.
13. Wilkinson, W.E., Salazar, M.K., Uhl, J.E., Koepsell, T.D., DeRoos, R.L., and Long, R.J. 1992. Occupational Injuries: A study of health care workers at a northwestern health science center and teaching hospital. *AAOHN Journal* 40(6): 287–293.
14. Valanis, B.G., Hartsberg, V., and Shortridge, L. 1987b. Antineoplastic drugs. Handle with care. *AAOHN Journal* 35: 487–492.
15. U.S. Department of Health and Human Services. 1991. Healthy People 2000. Washington, D.C.: GPO.
16. Jagger, J. 1994. A new opportunity to make the health care workplace safer. *Advances in Exposure Prevention* 1(1): 1–2.
17. Centers for Disease Control and Prevention (CDC). 1992. The second 100,000 cases of acquired immunodeficiency syndrome—United States, June 1981–December 1991. *Morbidity and Mortality Weekly Report* 41(2): 28–29.
18. Editorial. 1993. Transmission of hepatitis C via blood splash into conjunctiva. *Scandinavian Journal of Infectious Diseases* 25: 270–271.
19. Jagger, J., Cohen, M., and Blackwell, B. 1994. EPINet: a tool for surveillance of blood exposures in health care settings. *Essentials of Modern Hospital Safety* 3: 223–239.
20. Marcus, R.A., Tokars, J.I., Culver, P.S., and McKibben, D.M. 1991. Zidovudine use after occupational exposure to HIV-infected blood. Abstract No. 979. Presented at the 1991 31st Interscience Conference on Antimicrobial Agents and Chemotherapy (ICAAC), Chicago, IL.
21. Henderson, D.K., Fahey, B.J., Willy, M., Schmitt, J.M., Carey, K., Koziol, D.E., Lane, H.C., Redio, J., and Saah, A.J. 1990. Risk for occupational transmission of human immunodeficiency virus type 1 (HIV-1) associated with clinical exposures. *Annals of Internal Medicine* 113(10): 740–746.
22. Jagger, J. 1994. Report on blood drawing: Risky procedures, risky devices, risky job. *Advances in Exposure Prevention* 1(1): 4–9.
23. McEvoy, M., Porter, K., Mortimer, P., Simmons, N., and Shanson, D. 1987. Prospective study of clinical, laboratory, and ancillary staff with accidental exposures to blood or body fluids from patients infected with HIV. *British Medical Journal* 294: 1595–1598.
24. Wilkinson, W. 1987. Occupational injury at a midwestern health science center and teaching hospital. *AAOHN Journal* 35(8): 367–376.
25. Doan-Johnson, S. 1992. Taking a closer look at needle sticks. *Nursing 92* 22(8): 24, 27.
26. Jackson, M.M., Dechairo, D.C., and Gardner, D.F. 1986. Perceptions and beliefs of nursing and medical personnel about needle-handling practices and needlestick injuries. *American Journal of Infection Control* 14(1): 1–10.

27. Linnemann, C.C., Cannon, C., DeRonde, M., and Lanphear, B. 1991. Effect of educational programs, rigid sharps containers, and universal precautions on reported needlestick injuries in health care workers. *Infection Control and Hospital Epidemiology* 12(4): 214–219.

28. Willy, M.E., Dhillon, G.L., Loewen, N.L., Wesley, R.A., and Henderson, D.K. 1990. Adverse exposures and universal precautions practices among a group of highly exposed health professionals. *Infection Control and Hospital Epidemiology* 11(7): 351–356.

29. American Health Consultants (AHC). 1990. One third of needlesticks go unreported at hospital. *Hospital Infection Control* 17(18): 107.

30. Hamory, B.H. 1983. Underreporting of needlestick injuries in a university hospital. *American Journal of Infection Control* 11(5): 174–177.

31. Moorhouse, A., Bolen, R., Evans, J., Gilchrist, J., Hart, M., McKinlay, K., Midgely, L., Montada, M., Munro, A., Smith, R., Stewart, F., and Torresan, N. 1994. Needlestick injuries: the shock and the reality. *CINA: Official Journal of the Canadian Intravenous Nurses Association* 10(1): 14–18.

32. Gerberding, J.L. 1995. Management of occupational exposures to blood-borne viruses. *New England Journal of Medicine* 332(7): 444–449.

33. Centers for Disease Control and Prevention (CDC). 1990c. Guidelines for preventing the transmission of tuberculosis in health-care settings, with special focus on HIV-related issues. *Morbidity and Mortality Weekly Report* 39(RR-17), 1–29.

34. Centers for Disease Control and Prevention (CDC). 1994. *HIV/AIDS Surveillance Report* 5(19).

35. Jagger, J., Hunt, E.H., Brand-Elnagger, J., and Pearson, R.D. 1988. Rates of needlestick injury caused by various devices in a university hospital. *New England Journal of Medicine* 5: 284–288.

36. Becker, M.H., Janz, N.K., Band, J., Bartley, J., Snyder, M.B., and Gaynes, R.P. 1990. Noncompliance with universal precautions policy: Why do physicians and nurses recap needles? *American Journal of Infection Control* 17(46): 232–239.

37. Kopfer, A.M., and McGovern, P.M. 1993. Transmission of HIV via a needlestick injury: Practice recommendations and research implications. *AAOHN Journal* 41: 374–381.

38. English, J.F. 1992. Reported hospital needlestick injuries in relation to knowledge/skill, design, and management problems. *Infection Control and Hospital Epidemiology* 13(5): 259–264.

39. Not available.

40. van Wissen, K.A., and Siebers, R.W. 1993. Nurses' attitudes and concerns pertaining to HIV and AIDS. *Journal of Advanced Nursing* 18(6): 912–917.

41. Ribner, B.S., Landry, M.N., Gholson, G.L., and Linden, L.L. 1987. Impact of a rigid, puncture resistant container system upon needlestick injuries. *Infection Control* 8(2): 63–66.

42. Lettau, L.A. 1992. The A, B, C, D, and E of viral hepatitis: Spelling out the risks for health care workers. *Infection Control and Hospital Epidemiology* 13: 77–81.

43. Jeffers, L.J., Perez, G.O., deMedina, M.D., Ortiz-Interian, C.J., Schiff, E.R., Reddy, K.R., Jimenez, M., Bourgoignie, J.J., Vaamonde, C.A., Duncan, R., Houghton, M., Choo, G.L., and Kuo, G. 1990. Hepatitis C infection in two urban hemodialysis units. *Kidney International* 38: 320–322.

44. Centers for Disease Control and Prevention (CDC). 1988. Update: Universal precautions for prevention of transmission of human immunodeficiency virus, hepatitis B virus, and other bloodborne pathogens in health-care settings. *Morbidity and Mortality Weekly Report* 37(24): 377–388.

45. Drusin, L.M., Sohmer, M., Groshen, S.L., Spiritos, M.D., Senterfit, L.B., and Christianson, W.M. 1987. Nosocomial hepatitis A infection in a pediatric intensive care unit. *Archives of Diseases in Children* 62: 690–695.
46. Edgar, W.M. and Campbell, A.D. 1985. Nosocomial infection with hepatitis A. *Journal of Infections* 10: 43–47.
47. Centers for Disease Control and Prevention (CDC). 1989. Measles prevention: Recommendations of the immunization practices advisory committee (ACIP). *Morbidity and Mortality Weekly Report* 38: 1–18.
48. Centers for Disease Control and Prevention (CDC). 1984. Rubella prevention. *Morbidity and Mortality Weekly Report* 33: 301–323.
49. Greaves, W.L., Kaiser, A.B., Alford, R.H., and Schaffner, W. 1980. The problem of herpetic whitlow among hospital personnel. *Infection Control* 1: 381–385.
50. Centers for Disease Control and Prevention (CDC). 1983b. Guideline for infection control in hospital personnel. *Infection Control* 4(4): 326–349.
51. Beneson, A.S., Ed. 1995. *Control of Communicable Diseases in Man*, 15th ed., Washington, D.C.: American Public Health Association.
52. Sherertz, R.J., and Hampton, A.L. 1987. Chapter 13: Infection control aspects of hospital employee health. In: Wenzel, R.P., Ed., *Prevention and Control of Nosocomial Infection*. Baltimore: Wiliams and Wilkins, 175–204.
53. Balcarek, K.B., Bagley, R., Cloud, G.A., and Pass, R.F. 1990. Cytomegalovirus infection employees of a children's hospital: No evidence for increased risk associated with patient care. *American Medical Association Journal* 263(I): 840–844.
54. Centers for Disease Control and Prevention (CDC). 1990. Guidelines for preventing the transmission of tuberculosis in health-care settings, with special focus on HIV-related issues. *Morbidity and Mortality Weekly Report* 39(RR-17), 1–29.
55. Centers for Disease Control and Prevention (CDC). 1991c. Nosocomial transmission of multi-drug-resistant tuberculosis among HIV-infected persons—Florida and New York, 1988–1991. *Morbidity and Mortality Weekly Report* 40(34): 585–591.
56. Wilkinson, W.E., and Rubadue, C.L. 1986. Nursing life's guide to hazards on the job. Part IV: Sterilizing agents. *Nursing Life* 6: 43–45.
57. Mattia, M.A. 1983. Hazards in the hospital environment: The sterilants ethylene oxide and formaldehyde. *American Journal of Nursing* 83: 240–243.
58. Danielson, N.E. 1986. *Ethylene Oxide Use in Hospitals. A Manual for Health Care Personnel*, 2nd ed., 5–11, 179–196.
59. Rutala, W.A., and Hamory B.H. 1989. Expanding role of hospital epidemiology: Employee health—chemical exposure in the health care setting. *Infection Control and Hospital Epidemiology* 10(6): 261–266.
60. Rogers, B., and Emmett, E.A. 1987a. Handling antineoplastic agents: Urine mutagenicity in nurses. *Image: Journal of Nursing Scholarship* 19: 108–113.
61. Hirst, M., Levin, L., Mills, D.G., Tse, S., and White, D. 1984. Occupational exposure to cyclophosphamide. *Lancet* January 28: 186–188.
62. Waksvik, H., Klepp, O., and Brogger, A. 1981. Chromosome analyses of nurses handling cytostatic agents. *Cancer Treatment Reports* 65(7–8): 607–610.
63. Selevan, S.G., Lindbohm, M.L., Hornung, R.W., and Hemminki, K. 1985. A study of occupational exposure to antineoplastic drugs and fetal loss in nurses. *New England Journal of Medicine* 313(19): 1173–1178.
64. Christensen, C.J., Lemasters, G.K., and Wakeman, M.A. 1990. Work practices and policies of hospital pharmacists preparing antineoplastic agents. *Journal of Occupational Medicine* 32(6): 508–512.

65. Brodsky, J., and Cohen, E. 1985. Health experiences of operating room personnel. *Anesthesiology*, 68: 461–463.

66. National Institute for Occupational Safety and Health. Occupational exposure to waste anesthetic gases and vapors. Cincinnati, OH: DHEW.

67. McAbee, R. 1988. Nursing and back injuries. *AAOHN Journal* 36: 200–209.

68. Owen, B.D. 1989. The magnitude of low-back problems in nursing. *Western Journal of Nursing Research* 11(2): 234–242.

69. Arad, D., and Ryan, M. 1986. The incidence and prevalence in nurses of low back pain: a definitive survey exposes the hazards. *Australian Nurse Journal* 16: 44–48.

70. Carney, R.M. 1993. Protect your nursing athletes! *Nursing Management* 24(3): 69–71.

71. Harber, P., Billet, E., Gutowski, M., SooHoo, K., Lew, M., and Roman, A. 1985. Occupational low back pain in hospital nurses. *Journal of Occupational Medicine* 27(7): 518–524.

72. Jensen, R.C. 1987. Back injuries among nursing personnel: research needs and justifications. *Research in Nursing and Health* 10: 29–38.

73. Jorgensen, S., Hein, H.O., and Gyntelberg, F. 1994. Heavy lifting at work and risk of genital prolapse and herniated lumbar disc in assistant nurses. *Occupational Medicine* 44(1): 47–49.

74. Marchette, L., and Marchette, B. 1985. Back injury: A preventable occupational hazard. *Orthopaedic Nursing* 4(6): 25–29.

75. Owen, B.D., and Garg, A. 1991. Reducing risk for back injury in nursing personnel. *AAOHN Journal* 39(1): 24–33.

76. Feldstein, A., Valanis, B., Vollmer, W., Stevens, N., and Overton, C. 1993. The Back Injury Prevention Project pilot study: Assessing the effectiveness of back attack, an injury prevention program among nurses, aides, and orderlies. *Journal of Occupational Medicine* 35(2): 114–120.

77. Larese, F., and Fiorito, A. 1994. Musculoskeletal disorders in hospital nurses: A comparison between two hospitals. *Ergonomics* 37(7): 1205–1211.

78. Greenwood, J.G. 1986. Back injuries can be reduced with worker training, reinforcement. *Occupational Health and Safety* 55(5): 26–29.

79. Venning, P.J., Walter, S.D., and Stitt, L.W. 1987. Personal and job-related factors as determinants of incidence of back injuries among nursing personnel. *Journal of Occupational Medicine* 28(10): 820–825.

80. Newman, S., and Callaghan, C. 1993. Work-related back pain. *Occupational Health* 45: 201–205.

81. Neuberger, J.S., Kammerdiener, A.M., and Wood, C. 1988. Traumatic injuries among medical center employees. *AAOHN Journal* 36(8): 318–325.

82. Mott, P.E., Mann, F.C., McLoughlin, Q., and Warwick, D.P. In: *Shift Work: The Social, Psychological, and Physical Consequences*. Ann Arbor, MI: University of Michigan Press, 1985: 9–19, 311–314.

83. Czeisler, C.A., Moore-Ede, M.C., and Coleman R. 1982. Rotating shift work schedules that disrupt sleep are improved by applying circadian principles. *Science* 217: 460–463.

84. Jung, F. 1986. Shiftwork: Its effect on health performance and well-being. *AAOHN Journal* 34(4):161.

85. Moore-Ede, M.C., and Richardson, G.S. 1985. Medical implications of shiftwork, *Annual Review of Medicine* 36: 607–617.

86. Gold, D.R., Rogacz, S., Bock, N., Tosteson, T.D., Baum, T.M., Speizer, F.E., and Czeisler, C.A. 1992. Rotating shift work, sleep, and accidents related to sleepiness in hospital nurses. *American Journal of Public Health* 82(7): 1011–1014.

87. Todd, C., Robinson, G., and Reid, N. 1993. 12-hour shifts: Job satisfaction of nurses. *Journal of Nursing Management* 1(5): 215–220.

88. LaDou, J. 1990. *Occupational Medicine*. Norwalk, CT: Appleton and Lange.

89. Scott, A.J., and LaDou, J. 1990. Shiftwork: Effects on sleep and health with recommendations for medical surveillance and screening. *Occupational Mecidine State of the Art Reviews*: Vol. 5: Shiftwork: Philadelphia, PA: Hanley and Belfus, 273–300.

90. Tepas, D.I. 1982. Work/sleep time schedules and performance. In: Webb, W.B. Ed., *Biological Rhythms, Sleep and Performance*. New York: John Wiley and Sons, 175–200.

91. Smith, M.J., Colligan, M.J., and Tasto, D.L. 1982. Health and safety consequences of shift-work in the food processing industry. *Ergonomics* 25: 133–144.

92. Rix, G. 1987. Staff sickness and its relationships to violent incidents on a regional secure psychiatric unit. *Journal of Advanced Nursing* 12(2): 223–228.

93. Wasserberger, J., Ordog, G.J., Kolodny, M., and Allen, K. 1989. Violence in a community emergency room. *Archives of Emergency Medicine* 6: 266–269.

94. Lavoie, F., Carter, G.L., Danzl, D.F., and Berg, R.L. 1988. Emergency department violence in United States teaching hospitals. *Annals of Emergency Medicine* 17(11): 1227–1233.

95. Carmel, H., and Hunter, M. 1989. Staff injuries from inpatient violence. *Hospital and Community Psychiatry* 40(1): 41–46.

96. Jones, M.K. 1985. Patient violence report of 200 incidents. *Journal of Psychosocial Nursing and Mental Health Services* 23(6): 12–17.

97. Lipscomb, J.A., and Love, C.C. 1992. Violence toward health care workers: An emerging occupational hazard. *American Association of Occupational Health Nurses Journal* 40(5): 219–228.

98. Lanza, M.L. 1988. Factors relevant to patient assault. *Issues in Mental Health Nursing* 9: 239–257.

99. Poster, E.C., and Ryan, J.A. 1989. Nurses attitudes toward physical assaults by patients. *Archives of Psychiatric Nursing* 3(6): 315–322.

100. McCullough, L.E., McNeil, D.E., Binder, R.L., and Hatcher, C. 1986. Effects of a weapon screening procedure in a psychiatric emergency room. *Hospital and Community Psychiatry* 37: 837–838.

101. Lanza, M.L., and Milner, J. 1989. The dollar cost of patient assault. *Hospital and Community Psychiatry* 40(12): 1227–1229.

102. Lanza, M.L. 1983. The reactions of nursing staff to physical assault by a patient. *Hospital and Community Psychiatry* 34(1): 44–47.

103. Godwin, G.L., Bucholtz, J.D., and Wall, S.C. 1985. Hidden hazards on the job part III: Radiation. *Nursing Life* 5: 43–47.

104. Plog, B.A., Benjamin, G.S., and Kerwin, M.A. 1988. *Fundamentals of Industrial Hygiene*, 3rd ed., Chicago: National Safety Council.

105. Falk, S.A., and Wood, N.F. 1973. Hospital noise-levels and potential health hazards. *New England Journal of Medicine* 289: 774–781.

106. National Safety Council. 1988b. Accident prevention manual for industrial operations: Engineering and technology. 9th ed., Chicago: National Safety Council.

107. Levy, B., and Wegman, D. 1995. *Occupational Health*. Boston: Little, Brown and Company.

108. Doering, L. 1990. Recruitment and retention: Successful strategies in critical care. *Heart and Lung* 19(3): 220–224.

109. Fielding, J., and Weaver, S.M. 1994. A comparison of hospital- and community-based mental health nurses: Perceptions of their work environment and psychological health. *Journal of Advanced Nursing* 19(6): 1196–1204.

110. MacNeil, J.M., and Weisz, G.M. 1987. Critical care nursing stress: Another look. *Heart and Lung* 16(3): 274–277.

111. Hawley, M.P. 1992. Sources of stress for emergency nurses in four urban Canadian emergency departments. *Journal of Emergency Nursing* 18(3): 211–216.

112. Foxhall, M.J., Zimmerman, L., Standley, R., and Bene, B. 1990. A comparison of frequency and sources of nursing job stress perceived by intensive care, hospice, and medical-surgical nurses. *Journal of Advanced Nursing* 15(5): 577–584.

113. Boumans, N.P., and Landeweerd, J.A. 1994. Working in an intensive care or non-intensive care unit: Does it make any difference? *Heart and Lung* 23(1): 71–79.

114. Snape, J., and Cavanagh, S.J. 1993. Occupational stress in neurosurgical nursing. *Intensive and Critical Care Nursing* 9(3): 162–170.

115. Lewis, S.L., Campbell, M.A., Becktell, P.J., Cooper, C.L., Bonner, P.N., and Hunt, W.C. 1992. Work stress, burnout, and sense of coherence among dialysis nurses. *ANNA Journal* 19(6): 545–553.

116. Emery, J.E. 1993. Perceived sources of stress among pediatric oncology nurses. *Journal of Pediatric Oncology Nursing* 10(3): 87–92.

117. Gordon, J.H., Ulrich, C., Feeley, M., and Pollack, S. 1993. Staff distress among hemophilia nurses. *AIDS Care* 5(3): 359–367.

118. American Nurses' Association. 1984. *American Nurses: Association Task Force on Addictions and Psychological Dysfunctions.* Kansas City, MO.

119. Cull, A.M. 1991. Studying stress in care givers: Art or science? *British Journal of Cancer* 64(6): 981–984.

120. Hudson, G. 1990. The toxic ecology of work: Are the carers taking care? *Australian Nurse Journal* 19(10): 17.

121. Katz, R.M. 1983. Causes of death among registered nurses. *Journal of Occupational Medicine* 25: 760–762.

122. Keeve, J.P. 1984. Physicians at risk: Some epidemiologic considerations of alcoholism, drug abuse, and suicide. *Journal of Occupational Medicine* 26: 503–507.

123. Leppanen, R.A., and Olkinuora, M.A. 1987. Psychological stress experiences by health care personnel. *Scandinavian Journal of Work Environmental Health* 13: 1–8.

124. van Servellen, G., and Leake, B. 1994. Emotional exhaustion and distress among nurses: How important are AIDS-care specific factors? *Journal of the Association of Nurses in AIDS Care* 5(2): 11–19.

125. Wolfgang, A.P., Perri, M., and Wolfgang, C.F. 1988. Job-related stress experienced by hospital pharmacists and nurses. *American Journal of Hospital Pharmacy* 45(6): 1342–1345.

126. Bernstein, H.A. 1981. Survey of threats and assaults directed toward psychotherapists. *American Journal of Psychotherapy* 35(4): 542–549.

127. Cohen, S.R. 1987. Skin disease in health care workers. In Emmett, E.A. Ed., *Occupational Medicine. State of the Art Reviews* (Vol. 2 no. 3) *Health Problems of Health Care Workers.* Philadelphia: Hanley and Belfus, 565–580.

128. Gestal, J.J. 1987. Occupational hazards in hospitals: Risk of infection. *British Journal of Industrial Medicine* 44: 435–442.

129. Larkin, E., Murtagh, S., and Jones, S. 1988. A preliminary study of violent incidents in a special hospital (Rampton). *British Journal of Psychiatry* 153: 226–231.

130. Maslach, C., and Jackson, S.E. 1982. Burnout in health professions: A social psychological analysis. In Sanders, G. and Suls, J., Eds., *Social Psychology of Health and Illness*. Hillandale, NJ: Lawrence Erlbaum, 79–103.

131. Lewy, R.M. 1991. *Employees at Risk*. New York: Van Norstrand Reinhold.

132. Centers for Disease Control and Prevention (CDC). 1996. Disease burden from viral hepatitis A, B and C in the United States. CDC Fax Information Service.

133. Centers for Disease Control and Prevention (CDC). 1996. AIDS Information. Reported cases of AIDS and HIV infection in health care workers. CDC Fax Information Service.

134. Gurevich, I. 1994. Preventing needlestick: A market survey. *RN* 57(11): 44–49.

135. Makofsky, D., and Cone, J. 1993. Installing needlestick disposal boxes closer to the bedside reduces needle-recapping rates in hospital units. *Infection Control and Hospital Epidemiology* 14(3): 140–144.

136. Younger, B., Hunt, E., Robinson, C., and McKenmore, C. 1992. Impact of a shielded saftey syringe on needlestick injuries among health care workers. *Infection Control and Hospital Epidemiology* 13: 349–353.

137. Prince, K., Summers, L., and Knight, M.A. 1994. Needleless IV therapy: Comparing three systems for safety. *Nursing Management* 25(3): 80–83.

138. Atler, M.J. 1994. Occupational exposure to hepatitis C virus: A dilemna. *Infection Control and Hospital Epidemiology* 15: 742–744.

139. Atler, M.J. 1993. The detection, transmission, and outcome of hepatitis C virus infection. *Infectious Agents and Disease* 2: 155–166.

140. Centers for Disease Control and Prevention (CDC). 1994. Hepatitis surveillance, Report No. 55, 1–33.

141. Forseter, G., Wormer, G., Adler, S. et al. 1993. Hepatitis C in the health care setting II. Seroprevalence among hemodialysis staff and patients in suburban New York City. *American Journal of Infection Control* 21: 5–8.

142. Lanphear, B.P., Linnemann, C., Connon, C. et al. 1994. Hepatitis C virus infection in health care workers: Risk of exposure and infection. *Infection Control and Hospital Epidemiology* 15: 745–750.

143. Gerberding, J.L. 1994. Incidence and prevalence of human immunodeficiency virus, hepatitis B virus, hepatitis C virus, and cytomegalovirus among health care personnel at risk for blood exposure: Final report from a longitudinal study. *Journal of Infectious Diseases* 170: 1410–1417.

144. Centers for Disease Control and Prevention (CDC). 1992. Cytomegalovirus: Circumstances when CMV infection could be a problem. CDC Fax Information Service.

145. Garner, J.S. 1996. Guideline for isolation precautions in hospitals. Evaluation of isolation practices. *American Journal of Infection Control* 24: 24–52.

146. Wenger, P.N., Otten, J., Breeden, A., et al. 1995. Control of nosocomial transmission of multidrug-resistant *Mycobacterium tuberculosis* among health care workers and HIV-infected patients. *Lancet* 345: 235–240.

147. Centers for Disease Control and Prevention (CDC). 1996. Tuberculosis morbidity—United States, 1995. *Morbidity and Mortality Weekly Report* 45(18): 365–370.

148. Staff. 1993. APIC position paper. Prevention of device-medicated blood-borne infections to health care workers. *American Journal of Infection Control* 21(2): 76.

149. Charpin, D., Logier, R., Cheronet, I., and Vervolet, D. 1992. Prevalence of latex allergy in nurses working in operating rooms. *Journal of Allergy and Clinical Immunology* 87: 269.

150. Morales, C., Bosomba, A., Carriera, J., and Sastre, A. 1989. Anaphylaxis produced by rubber glove contact. *Clinical and Experimental Allergy* 19: 425–430.

151. Beezhold, D.M., Kostyal, D.A., and Wiseman, J. 1994. The transfer of protein allergies from latex gloves. *AORN Journal* 59(3): 605–613.
152. McCormack, B., Coveron, M., and Biel, L. 1995. Latex sensitivity: An occupational health strategic plan. *AAOHN Journal* 43(4): 190–196.
153. Alenius, H., Turjanmao, K., Makinen-Kiljunen, S., Reunala, T., and Polosuo, T. 1994. IgE immune response to rubber proteins in adult patients with latex allergy. *Journal of Allergy and Clinical Immunology* 93: 859–863.

3 The Epidemiology of Sharps Injuries

James E. Cone, Kim Hagadone, and Alisa Boyd

CONTENTS

INTRODUCTION

In a field poll of Californians, public concern about acquired immunodeficiency syndrome (AIDS) topped the list, with more than seven out of ten people extremely concerned about controlling the spread of AIDS. Since 1981, more than 53,000 people have died of AIDS in California alone. In terms of occupational infection with human immunodeficiency virus (HIV), as of June 1996, there were 51 definite and 108 possible cases of *reported* infection of health care workers in the United States. The true number is difficult to determine, however, due to documented frequent underreporting.

The transmission of AIDS and other bloodborne infectious diseases, including hepatitis B and hepatitis C, poses a serious threat to health care workers. Unlike tuberculosis (TB), which took decades of study before the risk to health care workers was widely accepted, infectious hepatitis and HIV infection were early recognized as risks of needlestick injuries. However, health care institutions have not readily embraced newer safety-enhanced medical devices as a method of prevention, citing the increased costs and unproved efficacy of these devices.

OBJECTIVES

- Review current data on the sharps injury epidemic among health care workers.
- Discuss current efforts to reduce injury through design of new sharps devices.
- Introduce the new California SHARPS program and present preliminary surveillance data on needlestick injuries in California.

BACKGROUND

Occupational exposure to bloodborne pathogens is a major concern among health care workers. The greatest risk for bloodborne pathogen transmission is associated with percutaneous injuries involving hollow-bore needles and other sharps objects contaminated with patient blood.[1] These injuries can occur when health care workers use, dissemble, or dispose of the sharps instrument. Housekeeping personnel are at risk for injury by sharps that, due to improper disposal, are concealed in linen and garbage. An estimated 800,000 to 1 million needlestick injuries are reported each year in the United States. With needlestick injuries accounting for approximately 80% of all accidental exposures to blood, occupational exposures from needles are a major concern. [2]

The increased awareness and recognized consequences of transmission of HIV, hepatitis B virus (HBV), and hepatitis C virus (HCV) from a needlestick injury have contributed to this concern. According to the Centers for Disease Control and Prevention (CDC) the estimated risk is 0.3% for seroconversion to HIV after a percutaneous exposure to blood from an HIV-infected patient. [4] The risk of acquiring HBV from a percutaneous injury with a contaminated needle is between 6 and 30%.[5] Estimates of the risk of transmission after percutaneous exposure to blood positive for anti-HCV antibody (anti-HCV) is 0–10%.[4] There is documentation that at least 20 different pathogens have been transmitted to humans from a needlestick injury.[3] Effective prevention programs, including the use of products designed to reduce exposure to blood and blood products in the health care setting, must be a major focus of our efforts.

CASE REPORT

The March 27, 1997 issue of the *New England Journal of Medicine* documents a simultaneously acquired infection with HIV and HCV from a single source.[4] The 48-year-old health care worker sustained a deep injury with a blood-contaminated needle while performing a phlebotomy on a patient with AIDS. At the time of the exposure the patient was not recognized as having HCV infection and had no clinical evidence of liver disease. In addition to the needlestick injury, blood from the collection tube also spilled into the spaces between the cuffs of the health care worker's gloves and wrists onto her chapped hands that were full of open cracks. The unusual aspect about this case is that the times to seroconversion were unusually long for both HIV and HCV. Seroconversion to HIV was detected between 8 and 9 months after exposure, and HCV seroconversion occurred between 9 and 13 months after exposure. The clinical course of the health care worker's illness was rapid, with hepatic coma and progressive renal failure and death 18 months after documented seroconversion and 28 months after the needlestick injury. The rapid course may have been related to the simultaneous acquisition of two infections. Although testing for the HIV antibody more than 6 months after exposure is not routinely recommended, if clinical signs and symptoms of infection persist for more than 6 months, evaluation for late seroconversion may be needed.

EPIDEMIOLOGY

McCormick et al.[6] conducted a 4-year epidemiological study of sharps injuries in health care workers in the pre-AIDS era (1975–1979) and in the current AIDS era (1987–1988), and reported those trends over the 14-year period. In the initial study it was determined that frequent injuries were related to needle disposal; therefore, the hospital needle disposal system was changed. Puncture-resistant disposal boxes were provided at each bedside and in each outpatient clinic. These changes have reduced injuries related to disposal 2.3-fold over 10 years. Some of the goals in the second AIDS study were to identify those personnel and job activities at highest risk of sharps injuries, and to develop guidelines for prevention.

By comparing the data from 1975–1979 with the 1987–1988 data, the annual incidence of reported sharps injuries tripled from 60.4/1000–187.0/1000 health care workers (Figure 1). A review of the data from the time period between 1987 and 1988 reveals that phlebotomists have the highest risk of sharps injuries with an annual rate of 407/1000 health care workers (Figure 2). Environmental personnel (including housekeeping) experienced the next highest incidence of injury with 305.8/1000 health care workers, followed by nursing personnel (196.5/1000 health care workers), and laboratory personnel (169.9/1000 health care workers).

According to the McCormick analysis, the activity most commonly associated with sharps injuries was collection and disposal of waste, linen, and procedure trays, which accounted for 19.7% of total injuries (Figure 3). Next, 16.0% of sharps injuries occurred in the operating room followed by 15.7% of sharps injuries during administration of parenteral injections or infusion therapy. Phlebotomy contributed to

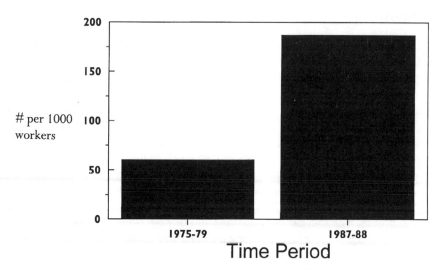

FIGURE 1   Annual incidence of needlesticks. (Data from McCormick RD et al. *Am J Med.* 1991;91:301S–307S.)

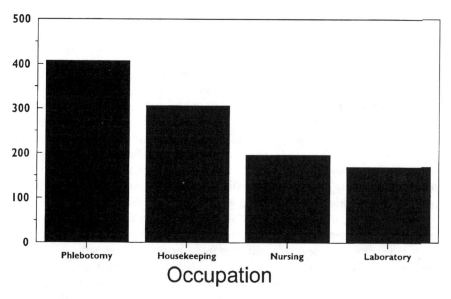

FIGURE 2   Proportion of needlesticks by occupation. (Data from McCormick RD et al. *Am J Med.* 1991;91:301S–307S.)

13.3% of sharps injuries, and 10.1% were due to recapping of used needles. Another 17.9% of sharps injuries are categorized as miscellaneous or indeterminate. Injuries associated with laboratory accidents account for 3.8%, and 3.5% of injuries occur during disposal of a used sharp device.

More data are available through the EpiNet data-sharing network coordinated by the International Health Care Worker Safety Center based at the University of

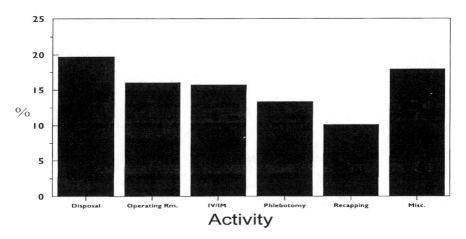

FIGURE 3 Proportion of needlesticks by activity. (Data from McCormick RD et al. *Am J Med.* 1991;91:301S–307S.)

Virginia located in Charlottesville, VA.[10] For 1996, there were 65 hospitals that contributed data. According to the Uniform Needlestick and Sharp-Object Injury Report for 1996, the total number of needlestick and sharps injuries reported among participating hospitals was 3167, excluding injuries occurring before use.

Further classification of sharps injury data was performed to promote an understanding of the distribution of sharps injuries. Some of the classification groups include job category, place of occurrence, original purpose of sharp item, how injury occurred, and device used (if a needle, whether it was safety enhanced). It is important for health care facilities to report and understand this information, especially when making decisions on which products to purchase.

JOB CATEGORY

Registered nurses (RNs) and licensed vocational nurses (LVNs) were the most common health care workers reporting a sharps injury with 1450 (46% of total) injuries reported in 1996. The second most common job experiencing a sharps injury was that of physicians (including attending/staff and intern/resident/fellow) who reported 475 (15%), followed by other workers with 278 injuries (9%). Surgery attendants, technologists (non-laboratory), and housekeeping/laundry workers reported 146, 148, and 150 sharps injuries, respectively, each accounting for 5% of the total reported sharps injuries.

ORIGINAL PURPOSE OF THE SHARP DEVICE

To identify potential high-risk procedures performed by health care workers, data were collected about the procedure being performed when the sharps injury occurred. The most commonly reported purpose of a sharp item was injection (intramuscular [IM]/subcutaneous), accounting for 492 (16%) of the injuries. Phlebotomy 421 (13%) and suturing 398 (13%) were the next commonly reported uses of sharp

TABLE 1
1996 EpiNet Data on Occupation

Occupation	Number	%
Nurse	1450	46
Physician	475	15
Other	278	9
Housekeeping/laundry worker	150	5
Technologist (nonlaboratory)	148	5
Surgery attendant	146	5
Phlebotomist/IV team/venipuncture	137	4
Clinical lab worker	123	4

TABLE 2
1996 EpiNet Purpose of Sharp Device

Purpose	Number	%
Injection, IM/sub-cutaneous	492	16
Drawing venous blood	421	13
Suturing	398	13
Other	370	12
Unknown	341	11
Cutting (surgery)	249	8
Starting IV insertion or heparin lock	218	7
Other injection/aspiration IV	156	5
Heparin or saline flushing	113	4

devices when the injury occurred. Other procedures (11%) and unknown procedures (12%) also accounted for a significant number of sharps injuries. Surgery also was a commonly reported procedure with 249 injuries (8%) occurring, and finally 218 injuries (7%) occurred when a health care worker started an intravenous (IV) insertion or heparin lock.

EVENT SURROUNDING SHARPS INJURY

Table 3 shows that 874 (28%) of the sharps injuries were reported to occur during use. The next most common time period when sharps injuries were reported was after use but before disposal, accounting for 701 sharps injuries (22%). Some 380 sharps injuries (12%) occurred between steps in a multistep procedure, while other unspecified events contributed 258 sharps injuries (8%). Finally, 219 sharps injuries (7%) were reported to occur while placing the sharp device into the disposal container, and injuries occurring during disassembling of the sharp device accounted for 190 injuries (6%).

TABLE 3
1996 EpiNet Event Surrounding Sharps Injuries

Event	Number	%
During use	874	28
Other after use, before disposal	701	22
Between steps	380	12
Other	258	8
Placing item into disposal container	219	7
Disassembling	190	6

RISK FACTORS FOR CONVERSION

To assess the potential risk factors associated with HIV seroconversion after percutaneous exposure to HIV-infected blood, a retrospective case control study was conducted by the CDC in collaboration with the French and British public health authorities.[7] Based on this study, risk factors that are associated with HIV transmission include deep injury, device being visibly contaminated with the source patient's blood, procedures involving a needle placed directly in the patient's vein or artery, and terminal illness in the source patient with death within 60 days postexposure and high viral titer. Identification of these risk factors suggests that the risk for HIV infection exceeds 0.3% after percutaneous exposures involving a larger volume of blood or higher HIV titer.[9]

CHANGING COSTS OF NEEDLESTICK INJURIES

The estimated cost per needlestick injury in 1990 was $405. In 1996, the estimated cost per needlestick injury for the first month, including post-exposure prophylaxis was $500–$800 plus similar amounts per month if additional prophylaxis was recommended.

PRELIMINARY ANALYSIS OF CALIFORNIA DOCTORS' FIRST REPORT DATA

METHODS

As part of a pilot surveillance system for California, all California doctor's first reports (DFRs) of work injury or illness for a 2-week period (May 12–26, 1997) were reviewed, and any that mentioned needlestick or sharps injury were selected. Included in the analysis were all non-duplicate cases that involved a needlestick or lancet injury. Other sharps-related injuries (e.g., knives, scissors, saw blades, and nail guns) were not included. Each case was classified with respect to age group, gender, occupation, industry, type of activity, type of device (if listed), and type of treatment and postexposure prophylaxis.

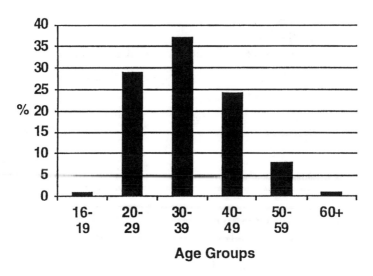

FIGURE 4 Age category—California DFRs.

RESULTS

There was a total of 93 needlestick/sharps injuries reported in California in the 2-week period sampled. The geographic distribution appeared skewed toward northern California, with only four reported cases from Los Angeles, compared with eight reported from Sacramento. Some major teaching hospitals did not report a single needlestick.

Age

The age was skewed toward younger age groups (Figure 4). The mean age was 35.29 years ± 10 years. The age range was from 17–62.

Gender

The majority of needlestick/sharps injuries were to women (59%).

Occupation

The largest single group reporting needlesticks was composed of nurses (28%), followed by technicians (14%), other (13%), and emergency responders (9%) (Table 4).

Industry

The most common industry where needlesticks occurred was that of hospitals (49%), followed by fire/police/emergency medical technician (EMT) departments (9%), schools/colleges (8%), and prisons (5%) (Table 5).

TABLE 4
California DFR Data on Sharps Injuries
by Occupation

Occupation	Number	%
Nurse	26	28
Technician	13	14
Other	12	13
Emergency responder	8	9
Dental	6	6
Physician	5	5
Housekeeping	5	5
Student	5	5
Not listed	5	5
Medical assistant	4	4
Phlebotomist	3	3
Child care worker	1	1
Total	93	

TABLE 5
California DFR Data on Sharps Injuries
by Industry

Industry	Number	%
Hospital	46	49.5
Fire/police/EMT	8	8.6
School/college	7	7.5
Other	7	7.5
Prison/jail	5	5.4
Dental offices	4	4.3
Home health agency	3	3.2
Medical clinic	3	3.2
Retail sales	3	3.2
Maintenance	3	3.2
Laundry	2	2.2
Manufacturing	1	1.1
Blood bank	1	1.1
Total	93	100.0

Event

The most common event associated with a sharps injury was phlebotomy (15%), followed by disposal of used sharps (14%), IV insertion, injection, and housekeeping activities (all 5%) (Table 6).

TABLE 6
California DFR Data on Event Surrounding Sharps Injuries

Event	Number	%
Phlebotomy	14	15.1
Disposal	13	14.0
IV insertion	5	5.4
Injection	5	5.4
Housekeeping	5	5.4
Suturing	3	3.2
Surgery	3	3.2
Laundry	3	3.2
Dental procedure	2	2.2
Recapping used needle	2	2.2
Cleanup after procedure	1	1.1
Altercation	1	1.1
Medication (clean needle)	1	1.1
Other	19	20.4
Not listed	16	17.2
Total	93	100.0

Type of Sharp Device Involved in Injury

Needles were the most common device involved (73%), followed by IV stylets (6%), suture needles (4%), and lancets (3%). Butterfly needles were specifically noted in 1% of reports. Otherwise, the specific type of needle was not listed in any reports.

Treatment Provided

The treatment provided after the sharps injury was most often not listed. For those treatments that were listed, the most common was tetanus toxoid injection (18%), followed by hepatitis B immunoglobulin (Ig, 10%) and post-exposure prophylaxis for HIV (2.2%).

DISCUSSION

This pilot study of sharps injuries reported on the DFRs in California was intended to evaluate the utility of this data source as a statewide surveillance tool. The primary weakness of this source is likely underreporting, which is most apparent when geographic distribution is evaluated. Certain hospitals apparently never report sharps injuries on DFRs, and certain regions seem to underreport significantly (e.g., Los Angeles). The strength of this data source is its statewide scope (with the problems noted). The lack of specific information on type of needle device makes this a less useful data source as well, although this might possibly be obtained by further interview or questionnaire to the patient or physician filling out the DFR.

SAFETY-ENHANCED MEDICAL DEVICES

Data indicate that there is no single solution for preventing needlestick injuries.[3] However, the greatest impact in reducing sharps injuries in health care workers might be achieved by innovative technology-based approaches to prevention.[6] The optimal solution is to reduce the use of needles by using alternative methods for performing medical procedures or whenever possible, eliminate needles from medical devices.[3] For instruments that require needles, the best approach is to design devices that allow the needle to remain covered during and after use. The worker's hands should remain behind the needle as it is covered to prevent movement in the direction of the used needle.[3] Hospitals could pay an additional 36% for preventive devices without increasing total costs.

THE CALIFORNIA SHARPS PROGRAM

The California Legislature, in response to public concern about needlesticks and other sharps hazards to health care workers, mandated, in its 1996 session, that a pilot surveillance study be undertaken. The multiyear study has been delegated to the state's department of health services (DHS). DHS has initiated the Sharps Injuries Surveillance and Prevention Program (SHARPS) to carry out the legislative mandate. The key feature of this study is the identification, by type and brand, of the sharp medical devices involved in health care workers' injuries. This involves monitoring sharps injuries in health care facilities in California, which include hospitals, skilled nursing facilities, and home health care agencies. The intended goal is to establish a record of the relative safety of each specific sharp medical device. The resulting data, scheduled to be available by year's end 1999, should allow health care institutions to benefit from each other's experiences and encourage medical device manufacturers to develop safer products.

REFERENCES

1. Hibberd PL. Patients, needles, and health care workers. *J Intravenous Nursing.* 1995;18:65–76.
2. Jagger J, Hunt E, Brand-Elnaggar R, Pearson R. Rates of needlestick caused by various devices in a university hospital. *N Engl J Med.* 1988;319:284–288.
3. Ridzon R, Gallagher K, Ciesielski C, Mast EE, Ginsberg MB, Robertson BJ, Luo CC, DeMaria, A. Simultaneous transmission of human immunodeficiency virus and hepatitis C virus from a needlestick injury. *N Engl J Med.*1997;336:919–922.
4. Special report and product review. Health Devices.
5. Aiken LH, Sloane DM, Klocinski MA. Hospital nurses occupational exposure to blood: prospective, retrospective, and institutional reports. *Am J Public Health.* 1997;87:103–107.
6. McCormick RD, Meisch MG, Ircink FG, Maki DG. Epidemiology of hospital sharps injuries: a 14-year prospective study in the pre-AIDS and AIDS eras. *Am J Med.* 1991;91:301S–307S.

7. Case-control study of HIV seroconversion in health care workers after percutaneous exposure to HIV-infected blood in France, United Kingdom, and United States, January 1988–August 1994. *MMWR*. 1995;44:929–933.
8. Gerberding JL. Management of occupational exposures to blood-borne viruses. *N Engl J Med*. 1995;332:444–451.
9. Notice to readers update: Provisional public health service recommendations for chemoprophylaxis after occupational exposure to HIV. *MMWR*. 1996;45:468–472.
10. Jagger J. Uniform Needlestick and Sharp-Object Injury Report 1996, 65 Hospitals. Adv Exposure Prev. 1997;3:15–16.

4 The Epidemic of Back Injuries in Health Care Workers in the United States

Bernice D. Owen

CONTENTS

STATEMENT OF THE PROBLEM

About 8 million workers are employed within the health care industry; this is 6.4% of the total United States workforce (Nelson and Olson, 1996). Even though these health care workers make up less than 10% of the workforce, they lead most other occupational groups in overexertion injury rates. The majority of these health care workers are nursing personnel working in a variety of settings within hospitals and also in community settings such as long-term care facilities, clinics, and patients' homes.

Back injuries have been a long-term significant problem for health care workers, especially nursing personnel who provide direct patient/resident care. Klein and co-workers (1984) found, through analysis of national workers' compensation claims, that nursing personnel ranked fifth among all workers for occupationally related back problems. Nursing personnel were surpassed only by very heavy physical labor groups that included miscellaneous laborers, sanitation workers, warehouse workers, and mechanics.

0-8493-3382-2/99/$0.00+$.50
© 1999 by CRC Press LLC

Part of the reason for this current epidemic in back injuries in health care workers may well be related to the changes that have occurred in the delivery of health care services. The health care system has changed drastically over the past few years. Patients are staying a much shorter time in the hospital; therefore, they are not in that setting during convalescence, which is generally a period when they need less physically stressful care. Many of these patients are going to long-term care and rehabilitation settings needing more care than patients in those settings have needed in the past; likewise, they are also returning home and needing more care than previously needed in that setting. The trend for hospice care is increasing so more patients are at home needing much care. These changes have placed a heavy burden of physically stressful work on health care workers in all these settings.

In 1990, the U.S. Department of Health and Human Services, Public Health Service, published the national health goals for the year 2000. The Occupational Safety and Health Objective (10.2) aims at reducing work-related injuries resulting in medical treatment, lost time from work, or restricted work activity to no more than six cases per 100 full-time workers. Because the rate is so high for nursing and personal health care workers (16.9), however, the national objective goal for this group is to decrease the number of work-related injuries to no more than nine per 100 full-time nursing and personal health care workers (American Public Health Association, 1990).

In private industry in 1993, there was an overall incidence rate of 8.5 cases of nonfatal workplace injuries and illnesses for every 100 equivalent of full-time workers. However, when considering only those industries with 100,000 or more of these injury cases, there was an incidence rate of 16.9 for every 100 full-time equivalent (FTE) workers for nursing and personal care facilities. This rate was surpassed only by the motor vehicle and equipment manufacturing industries (17.7). The next highest was attributed to the trucking and courier services (except air) with an incidence of 13.7 per 100 FTE workers (Bureau of Labor Statistics [BLS], 1994a).

Almost 50% of these nonfatal workplace injuries were serious sprains and strains, especially to the back. However in nursing, 67% of the disabling conditions were due to sprains and strains (BLS, 1994b). In fact, this same report indicated "these relatively hazardous occupations include ... female-dominated activities such as nursing care ..." (BLS, 1994b, p. 1). In the analysis of these nonfatal workplace injuries for nursing, the report indicated that 88% of the nursing assistants injured were female, 67% of the disabling conditions were due to sprains and strains, the disabling event was overexertion in 59% of the cases, and the source was patients in 61% of the cases. About one fifth of these sprain and strain injuries resulted in 6 weeks or more per case in absence from the job.

The profile of injuries and illnesses developed in 1992 by the BLS (1994b) confirmed the fact that overexertion injuries were high in occupational settings. They found sprains and strains were, by far, the leading category under nature of injury or illness; the back and other portions of the trunk were the leading categories of major body parts affected; and overexertion from lifting, pulling, or pushing heavy

---

1993 Statistics

Private Industry: 81 per 10,000 Workers

Nursing Home Personnel: 363 per 10,000 Workers

(Bureau of Labor Statistics, 1995)

1995 Statistics

Private Industry: 69 per 10,000 Workers

Nursing Home Personnel: 320 per 10,000 Workers

(Bureau of Labor Statistics, 1997c)

FIGURE 1   Overexertion injury rates for nursing home personnel vs. private industry.

or unwieldy objects or persons (often residents/patients in health care settings) led all other events or exposures. Based on the findings of this profile, one can assume that the majority of the injury cases (with the preceding incidence rate of 16.9 for workers in nursing and personal care facilities) were overexertion injuries to the back or trunk from lifting patients/residents.

The 1993 statistics indicated that nursing home personnel had four times more overexertion injuries than workers in private industry. In private industry, there were 81 reported overexertion injuries per 10,000 workers; in nursing there were 363 per 10,000 workers (BLS, 1995) (see Figure 1). Nursing assistants and orderlies lost a median of 6 days away from work but almost one fourth of them lost 3 weeks or more from work.

NURSING HOMES

Statistics for this same year (1993) indicated that nursing assistants in nursing homes led all other industries in these overexertion injuries (BLS, 1995).

In 1995, nursing home personnel continued to rank first for overexertion injuries with a rate of 320 lost work time cases per 10,000 workers. This rate is actually lower than in the past few years, but it continues to be four times greater than the 1995 national rate for overexertion injuries, which were 69 cases per 10,000 workers (BLS, 1997c) (see Figure 2).

1993 Injury Rates

Nursing Assistants in Nursing Homes Lead ALL Other Industries in Overexertion Injuries.

(Bureau of Labor Statistics, 1995)

1995 Injury Rates

Nursing Assistants in Nursing Homes Continue to Lead ALL Other Industries in Overexertion Injuries.

(Bureau of Labor Statistics, 1997c)

FIGURE 2   Occupation with highest incidence of overexertion injuries.

HOME CARE

Myers and colleagues (1993) did one of the first studies describing the back injury problem with home health care workers. They compared the back injury incident reports of hospital nursing assistants with nursing assistants employed in home care settings. They found the rate of back injuries significantly higher for those working in home care (5.9 per 100 FTE workers for hospital nursing assistants and 15.4 per 100 FTE workers in home care). For both groups, the majority of injuries occurred while carrying out patient-handling tasks, which were planned, vs. spontaneous tasks, such as catching a falling patient. Of the injuries for both groups, 40% occurred at the patient's bedside and included moving the patient up in bed. In addition, in 88% of the injuries in home care, the nursing assistant was working alone; 39% occurred in the hospital while working alone. At least 75% of both groups used no lifting/transferring equipment.

In home health care, the workforce doubled between 1989 and 1994 (from 250,000–500,000); it is estimated that this number will more than double by the year 2005 (BLS, 1997a). In 1994, the overexertion injury rates in these workers were 183 per 10,000 workers; this is more than double the national rate. This rate is even higher than the rate of hospital workers, which is 144 per 10,000 workers (see Figure 3). The average lost work days per injury in home health care workers is 7, which is 2 days more than health care workers in other settings (BLS, 1997b) (see Figure 4).

The studies from the BLS, (1997a) also indicated that the leading cause of injury for home health care workers was overexertion resulting from assisting patients, typically, through attempts at moving patients without assistance from other workers. These injuries accounted for nearly 40% of the industry's total 18,800 lost work

---

1994 Statistics

Home Health Care Workers: 183 per 10,000 Workers

Hospital Health Care Workers: 144 per 10,000 Workers

(Bureau of Labor Statistics, 1997b)

FIGURE 3   Overexertion injury rates for home health care workers vs. hospital workers.

Lost Work Days:

The Average Is 7 Days for Home Health Care Workers.

This Is 2 More Days than Health Care Workers in Other Settings.

(Bureau of Labor Statistics, 1997b)

FIGURE 4   Lost work days from overexertion injuries for home health care workers.

time injuries in 1994; the next highest number of injuries were those resulting from highway accidents, which were 16% of the injuries.

CONTRIBUTING FACTORS

LIFTING AND TRANSFERRING PATIENTS/RESIDENTS

Over the years many researchers and health care workers have implicated the task of lifting and transferring patients and residents as the major precipitating factor leading to these overexertion injuries (e.g., Bell, 1984; Harber et al., 1985; Jensen, 1985; Owen, 1987; Owen and Garg, 1993; Personick, 1990; Stubbs et al., 1983).

Owen and Garg (1993) and Owen and co-workers (1995) studied the back injury situation in nursing home and hospital settings. They, too, found the majority of reported injuries occurred while the health care worker (often working alone) lifted

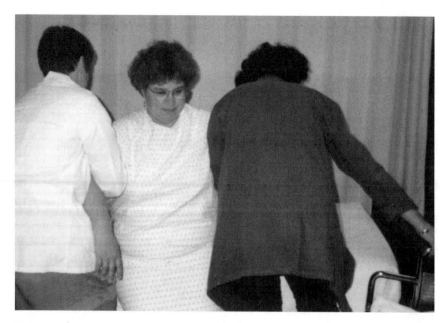

FIGURE 5   The technique used most frequently for the lifting and transferring of patients or residents.

or transferred patients at the bedside of the patient and used no equipment or assistive devices such as a gait belt or mechanical lift. Through observation they found the manual technique used most often in lifting and transferring patients involved grasping the patients under the axilla (armpit) and vertically lifting them to a standing position or lifting and carrying them to a nearby location (see Figure 5).

A major link to this problem of overexertion injuries is the manner in which patients are lifted and transferred. The most frequently observed method for lifting and transferring patients from one location to another is a manual lift carried out in the following manner: the nurse faces the patient, stoops and leans toward the patient, grasps the patient by placing his or her arm under the axilla of the patient, and then vertically lifts the patient to a standing position or lifts and carries the patient to a new location (see Figure 5). This technique might be carried out with one nurse or with one nurse on each side of the patient. This technique has been studied by Owen and Garg (1993) and Knibbe and Knibbe (1990). In both of these studies, the findings indicated the compressive force on L_5S_1 (the most vulnerable area of the low back) was much greater than L_5S_1 levels recommended by the National Institute for Safety and Health (U.S. Department of Health and Human Services, 1981). Many principles of body mechanics are in violation: the back is twisted, the lift is one-sided (asymmetrical), and the weight to be lifted exceeds recommendations. Therefore, when carrying out this lift, the nurse or nurses are very vulnerable to the sprain and strain injuries of the back.

There is evidence that even student nurses succumb to back pain problems related to the lifting and transferring of patients. Klaber-Moffett and colleagues (1993)

Professional Nurses	+ 65
Licensed Practical Nurses	50
Nursing Assistants	51
Other Professionals	85
Assistive Personnel	116
(Shogren et al., 1996)	

FIGURE 6   Percentage of increase in injury and illness rates since downsizing number of staff in 12 hospitals in Minnesota.

studied student nurses over a 30-month period and found that 64% reported at least a 1-day episode of back pain related to the lifting and transferring of patients; they also found that 37% of these students reported at least 3 consecutive days of back pain related to these same tasks.

REORGANIZATION AND RESTRUCTURING

With present downsizing and restructuring within many health care facilities, there are fewer nursing personnel and other health care workers available to care for patients. Patient acuity level is up and therefore patients need more physical care, yet there are fewer professional staff to care for these patients and fewer to guide the less skilled in care for patients.

The Minnesota Nurses Association conducted a 12-hospital study in 1995 and found a 65.2% increase in the number of injuries and illnesses among nurses between 1990 and 1994. During this same period, the number of nurses employed was reduced by 9.2%. At these same hospitals, the illness and injury rates for licensed practical nurses (LPNs) rose 50%, for nursing assistants went up 51%, for other professionals rose by 85.4%, and for assistive personnel went up by 116.6%. The majority of these injuries were related to moving patients (back, shoulder, and neck injuries) or to using sharps devices. This study and other studies have concluded there is a link between staffing levels and injuries (Shogren et al., 1996) (see Figure 6).

COSTS

The costs associated with the treatment of back injuries are very high. It is estimated that $50 billion per year are spent on such treatment (Borenstein, 1995; Johnson et

---

Thirty-eight Percent of Nurses Had Occupationally Related Back Pain.

Thirty-three Percent of These Reported It.

FIGURE 7   Percentage of nurses who had back pain and percentage who reported it.

al., 1996). These are only a portion of the real cost. Other costs for the worker include lost time from work, increased emotional and physical distress, job dissatisfaction, job and career changes, and role changes at home. There are also hidden costs for the institution, which include lost income due to loss in productivity, job turnover, orientation of new staff, staff morale, and administrative time for investigation and paperwork.

In a Taiwan hospital employing 998 nurses, almost half of these nurses indicated they experienced occupationally related back pain; 106 incurred economic costs related to this problem. The total monthly costs over a 12-month period ranged from US$105,405 to $149,083 (Lin et al., 1996).

TIP OF THE ICEBERG

There is no doubt that statistics indicate that back pain and compensated back injuries are a major problem for health care workers. There is some indication that these statistics may be revealing only the tip of the iceberg. In a random sampling of nurses, Owen (1989) found that 38% stated they had suffered at least three consecutive days of occupationally related back pain, but only 33% of that number ever reported this; Owen also found that 12% of those stating they had back pain were thinking of leaving nursing because of the back pain (see Figure 7).

SUMMARY

Health care workers in all settings within the health care system experience very high rates of overexertion injuries, especially to the back. In fact, nursing assistants in nursing homes lead all other industries in these overexertion injuries. Major contributing factors point to the manner in which patients and residents are lifted and transferred and also to the downsizing of staffing levels of health care workers in many health care settings. Costs to workers and health care industries are very high. There is some evidence that even though these rates are high, they may be even higher because many health care workers do not report their injuries.

REFERENCES

American Public Health Association. (1990). *Healthy People 2000: National Health Promotion and Disease Prevention Objectives*. Washington, D.C.: Author/Professional Affairs Division.

Bell, F. (1984). *Patient Lifting Devices in Hospitals*. London: Groom Helm.

Borenstein, D. (1995). Epidemiology, etiology, diagnostic evaluation, and treatment of low back pain. *Current Opinion in Rheumatology*, 8, 124–129.

Bureau of Labor Statistics (BLS)/U.S. Department of Labor. (1994a). *News* (USDL-94-600). Washington, D.C.: Author.

Bureau of Labor Statistics (BLS)/U.S. Department of Labor. (1994b). *News* (USDL-94-213). Washington, D.C.: Author.

Bureau of Labor Statistics (BLS)/U.S. Department of Labor. (1995). *News* (USDL-95-142). Washington, D.C.: Author.

Bureau of Labor Statistics (BLS)/U.S. Department of Labor. (1997a). Injuries to caregivers working in patients' homes. *Issues in Labor Statistics*, (Summary 97–94).

Bureau of Labor Statistics (BLS)/U.S. Department of Labor/Office of Safety, Health and Working Conditions. (1997b). Home health care services: injuries resulting in absences from work. *Lost Worktime Injuries and Work Hazards* (Supplement of News Release, USDL-96-163)

Bureau of Labor Statistics (BLS)/U.S. Department of Labor. (1997c). *News* (USDL 97-188). Washington, D.C.: Author.

Harber, P., Billet, E., Gutowski, M., SooHoo, K., and Roman, A. (1985). Occupational low-back pain in hospital nurses. *Journal of Occupational Medicine*, 27, 518–524.

Jensen, R. (1985). Events that trigger disabling back pain among nurses. *Proceedings of the 29th Annual Meeting of the Human Factor Society*. Santa Monica, CA.

Johnson, W.G., Baldwin, M.L., and Burton, J.F. (1996). Why is the treatment of work-related injuries so costly? New evidence from California. *Inquiry*, 33, 53–65.

Klaber-Moffett, J.A., Hughes, G.I., and Griffiths, P. (1993). A longitudinal study of low back pain in student nurses. *International Journal of Nursing Studies*, 30, 197–212.

Klein, B., Jensen, R., and Sanderson, L. (1984). Assessment of workers' compensation claims for back strains/sprains. *Journal of Occupational Medicine*, 26(6), 443–448.

Knibbe, J.J. and Knibbe, N.E. (1990). The workload on the back during the transfer from the wheelchair to the toilet, *Locomotion*, pp. 1–10. The Netherlands: Zwaag.

Lin, M., Tsauo, J., and Wang, J. (1996). Determinants of economic cost related to low back pain among nurses at a university hospital. *International Journal of Occupational and Environmental Health*, 2, 257–263.

Myers, A., Jensen, R., Nestor, D., and Rattiner, J. (1993). Low back injuries among home health aides compared with hospital nursing aides. *Home Health Care Services Quarterly*, 14(2/3), 149–155.

Nelson, M., and Olson, D. (1996). Health care worker incidents reported in a rural health care facility. *AAOHN Journal*, 44(3), 115–122.

Owen, B.D. (1987). The need for application of ergonomic principles in nursing. In S. Asfour (Ed). *Trends in Ergonomics/Human Factors IV*, pp. 831–838. Amsterdam: Elsevier Science Publishers.

Owen, B.D. (1989). The magnitude of low back problems in nursing. *Western Journal of Nursing Research*, 11(2), 234–242.

Owen, B.D., and Garg, A. (1993). Back stress isn't part of the job. *American Journal of Nursing*, 93(2), 48–51.

Owen, B.D., Keene, K., Olson, S., and Garg, A. (1995). An ergonomic approach to reducing back stress while carrying out patient handling tasks with a hospitalized patient. In Hagberg, Hofmann, Stobel, and Westlander. *Occupational Health for Health Care Workers,* pp. 298–301. Landsberg, Germany: ECOMED.

Personick, M.E. (1990). Nursing home aids experience in serious injuries. *Monthly Labor Review,* 113(2), 113–137.

Shogren, E., Calkins, A., and Wilburn, S. (1996). Restructuring may be hazardous to your health. *American Journal of Nursing,* 96(11), 64–66.

Stubbs, D., Buckle, P., Hudson, M., Rivers, P., and Worringham, C. (1983). Back pain in the nursing profession. I. Epidemiology and pilot methodology. *Ergonomics,* 26, 755–765.

U.S. Department of Health and Human Services. (1981). *Work practices guide for manual lifting* (DHHS/NIOSH, No. 81-122). Cincinnati, OH: Author.

5 Occupational Exposure to Tuberculosis*

CONTENTS

* Reprinted with permission from *Occupational Safety and Health Reporter*, Vol. 27, No. 21, pp. 667 et seq. (October 22, 1997). Copyright 1997 by The Bureau of National Affairs, Inc. (800-372-1033) <http://www.bna.com>.

OCCUPATIONAL EXPOSURE

Exposure to TB in the health care setting has long been considered an occupational hazard. With the steady decline in reported TB cases from 1953 to 1985, some of the concern for occupational exposure and transmission also declined. However, from 1985 to 1992 the number of reported cases of TB increased. In addition, in recent years, several outbreaks of TB among both patients and staff in hospital settings have been reported to the CDC. These outbreaks have been attributed to several factors: (1) Delayed recognition of active TB cases, (2) delayed drug susceptibility testing, (3) inadequate isolation of individuals with active TB (e.g., lack of negative pressure ventilation in isolation rooms, recirculation of unfiltered air, and allowing infectious patients to freely move in and out of isolation rooms), and (4) performance of high-risk procedures on infectious individuals under uncontrolled conditions (Ex. 7-50). In addition to hospitals, outbreaks of TB have also been reported among the patients, clients, residents and staff of correctional facilities, drug treatment centers, homeless shelters and long-term health care facilities for the elderly. The factors contributing to the outbreaks in these other occupational settings are very similar to those factors contributing to the outbreaks in hospital settings (i.e., delayed recognition of TB cases and poor/inadequate ventilation for isolation areas).

The following is a discussion of some of the studies that have examined occupational transmission of TB. A large proportion of the available information comes from exposures occurring in hospitals, in part because this occupational setting has been recognized for many years as an area of concern with regards to the transmission of TB. However, in more recent years this concern has spread to other occupational settings which share factors identified in the hospital setting as contributing to the transmission of disease. The following sections will include a discussion of some of the historical data from the hospital setting, as well as the more recent data that have been developed in hospitals and other occupational settings where the transmission of TB has occurred as a result of the recent resurgences in the number of active TB cases.

Hospitals—Prior to 1985

Even prior to the recent resurgence of TB in the general population, studies have shown an increased risk of transmission of TB to health care workers exposed to individuals with infectious TB. These studies clearly demonstrate that in the absence of appropriate TB control measures (e.g., lack of early identification procedures, lack of appropriate engineering controls), employees exposed to individuals with infectious TB have become infected and in some cases have developed active disease.

In 1979, Barrett-Connor (Ex. 5-11) examined the incidence of TB among currently practicing physicians who graduated from California medical schools from approximately 1950 to 1979. Through mailed questionnaires, physicians were asked to provide information that included their year of graduation from medical school, BCG vaccination history, history of active TB, results of their tuberculin skin testing, and the number of patients they were exposed to with active TB within the past year. They were also asked to classify themselves as tuberculin positive or negative and to indicate the year of the last negative and first positive tuberculin test.

Of the 6425 questionnaires mailed out, 4140 responses were received from currently practicing physicians. Twelve percent of the physicians had received the BCG vaccine. Sixty-one percent of the unimmunized physicians, who also had no history of active tuberculosis, considered themselves to be tuberculin negative. A total of 1542 (42%) reported themselves as having a positive response to the tuberculin skin test, with approximately 44 percent of those tuberculosis infections occurring before entering medical school. Of those infections occurring before entering medical school, approximately eight percent were reported as having been a result of contact following work experience in the hospital prior to entering medical school. For those physicians infected either during or after medical school, the sources of infection were reported as occurring as a result of a known patient contact (45.1%), an unknown contact (41.5%) and a non-patient contact (13.4%). In some cases, the nonpatient contact was reported as another physician or another hospital employee. Approximately one in ten of the physicians infected after entry into medical school developed active TB disease.

The authors also examined the incidence of infection, measured as the conversion rates in those remaining negative at the end of different time intervals (e.g., the last three years of medical school and five to 10 years after graduation). This examination indicated that from 1950 to 1975, there was a 78% decrease in tuberculin conversion rates despite the expanding pool of susceptible medical students (i.e., an increasing number of medical students who were tuberculin negative). Yet despite this overall decrease in infection rates over a 25 year period, tuberculin conversion rates among recent graduates exceeded 1% per year and age-specific infection rates among all the physicians studied were more than twice that of the U.S. population at comparable ages. The authors did not obtain information from the physicians on what type of infection control measures were being used in the facilities where they acquired their infections.

A similar analysis by Geisleler *et al.* (Ex. 7-46) evaluated the occurrence of active tuberculosis among physicians graduating from the University of Illinois medical school between the years 1938 and 1981. This study, also conducted by questionnaire, reported that among 4575 physicians questioned, there were 66 cases of active TB, of which 23% occurred after 1970. Sixty-six percent of the cases occurred within 6 years of graduation. In addition, the authors reported that in most years the incidence of TB was greater among these physicians than the general population.

Weiss (Ex. 7-45) examined tuberculosis among student health nurses in a Philadelphia hospital. From 1935 to 1939, before the introduction of anti-TB drugs and the beginning of the general decline of TB in the United States, 100% conversion rates were observed among those students who were initially tuberculin negative. For example, of 643 students admitted, 43% were tuberculin negative. At the end of only 4 months, 48% were tuberculin positive. At the end of 1 year, 85.9% were tuberculin positive and by the end of the third year 100% were positive. Of those students who converted during their student nursing tenure, approximately 5 percent developed active TB disease.

A decline in the rate of infection was observed over the next 36 years among student nurses at this hospital. The rates of infection were followed for ten classes of student nurses from 1962 to 1971. The students had little contact with patients during their first year but spent 4 weeks of their second year of training on the tuberculosis wards. Among those students initially tuberculin negative, the average conversion rate was 4.2% over the nine year period, ranging from 0 to 10.2%. Of the students who converted, 0.6% developed active TB disease. The authors attributed the decreases in conversion rates to not only the general decrease in TB disease in the community, but also to the increased efficiency of surveillance of patients entering the hospital for the early identification of potential cases of TB and the increased efficiency of isolation for TB patients. Despite the dramatic decreases in conversion rates among these student nurses, conversion rates were observed at levels as high as 10% for a given year, indicating that while the infection rates had decreased substantially since 1939, there still remained a significant amount of occupational transmission of TB in 1971. Moreover, this study shows that short term exposure, i.e., 4 weeks, is capable of infecting hospital employees.

Similar rates of conversion among hospital employees initially tuberculin negative were observed in a 1977 study by Ruben *et al.* (Ex. 7-43) which analyzed the results of a tuberculin skin testing program 31 months after its inception at a university hospital in Pittsburgh. Of 626 employees who were tested twice with the tuberculin skin test, 28 (4.5%) converted from negative to positive. The employees were classified as either having a "presumed high degree of patient exposure" or a "presumed low degree of patient exposure". Employees presumed to have high patient exposure included nurses, X-ray and isotope laboratory personnel and central escort workers. Employees presumed to have low exposure included secretaries, persons in housekeeping and dietary work, and business office, laundry and central supply personnel. The rates of conversion for employees with presumed high exposure (6%) and for employees with presumed low exposure (8%) were not significantly different. However, this study excluded physicians and medical and nursing students. These groups of employees would also presumably have had high exposure to patients since they are often the hospital staff most directly involved in administering patient care. Had these employees been included the number of conversions among employees with presumably high exposure may have been significantly increased.

The study was not designed to determine the source of exposure for any of the employees who converted. However, the authors suggested that the high level of conversions among those employees with presumed low exposure to patients may have resulted from exposures at home. A majority of this group was comprised of housekeeping staff who were of low socio-economic status. The authors also suggested that unrecognized cases of tuberculosis may be playing an important role in the occupational transmission of TB in the hospital.

Unrecognized cases of TB have been shown to play a significant role in the outbreak of TB in a general hospital. In 1972, Ehrenkranz and Kicklighter (Ex. 5-15) reported a case study in which 23 employees converted after exposure to a patient with an undetected case of tuberculosis bronchopneumonia. In this study, the source case was an individual who was admitted to the emergency room with pulmonary edema. Upper lobe changes of the lung were noted in the chest X-ray, and TB was mentioned as a possible cause. However, no sputum cytology was conducted. The patient spent 3 hours in the emergency room, 57 hours in a private room and another 67 hours in intensive care until his death. Treatment of the patient included intubation with an endotracheal tube and vigorous nasotracheal suctioning. It was only upon microscopic examination of tissue samples of the lung and lymph nodes after the autopsy of the patient that tuberculosis mycobacteria were detected.

Employees who worked in the emergency room, the intensive care unit and on the floor of the private room (NW 3) and who were also tuberculin negative before the admission of the patient, were retested to detect possible conversion. In addition, 21 initially tuberculin negative employees on an adjacent floor (NW 2) were also retested. Of the 121 employees tested, 24 were identified as having converted to positive status (21 working on NW 3, 2 working in the intensive care unit and 1 working on NW 2). No conversions were observed among those working in the emergency room.

The employees who were retested were classified as either having close contact (e.g., providing direct care), little contact (e.g., more distant contact), unknown contact (e.g., no record or recollection of contact) or indirect contact (e.g., in the same room a day or two after the patient's stay). Conversions occurred in 50% (13 of 26) of those employees with close contact, 18.5% (6 of 33) of those with little contact, 21.4% (3 of 14) of those with unknown contact and 3.7% (1 of 29) of those with indirect contact.

While the majority of conversions seems to have occurred in those employees on NW 3 who had close or little contact, there also were employees with more distant contact who were infected. An analysis of the ventilation of NW 3 indicated that the central air conditioning recycled 70% of the air with no high efficiency filter and no record of balancing the air conditioning system, thus allowing the air from the patients' rooms to mix with and return to the central corridor air. In addition, smoke tube tests detected direct air flow from the patients' rooms to the hall corridor. Perhaps the more important factor was that the patient was not diagnosed with infectious TB until after his death, by which time he had already infected 24 employees.

These earlier studies illustrate that despite the decrease in TB morbidity since the advent of anti-tuberculosis drugs in the 1940's, occupational transmission of TB continues to be a problem. In addition, while many improvements have been made in infection control procedures for TB in hospitals, evidence of occupational transmission of TB continues to be reported.

Hospitals—1985 to Present

As discussed above, the transmission of TB has been well established as an occupational hazard in the hospital setting. Many improvements were made in infection control practices. However, the resurgence in TB from 1985 to 1992 has brought to attention the fact that many TB control measures have not been implemented or have been inadequately applied. These studies demonstrate that TB continues to be an occupational hazard in the hospital setting. In addition, similar to the earlier studies, the more recent data show that the lack of early identification procedures and the lack of appropriate ventilation, performance of high-hazard procedures under uncontrolled conditions and the lack of appropriate respiratory protection have resulted in the infection of employees and in some cases the development of active disease. The more current outbreaks are even more troubling due to the emergence of multidrug-resistant forms of TB disease, which in some cases have resulted in fatality rates approaching 75%.

In a 1985 study, Chan and Tabak (Ex. 7-3) investigated the risk of TB infection among physicians in training at a Miami hospital. In this study a survey was conducted among 665 physicians in training who were in their first four years of postgraduate training. Only 404 responded to the survey, of which 13 were illegible. Another 72 were excluded because they had received the BCG vaccination. Of the remaining 319 physicians, 55 were tuberculin positive.

Of the 279 who were tuberculin negative at the beginning of their post graduate training, 15 were excluded because they had more than four years of training and 43 were excluded because they had not had repeat skin tests. Of the 221 remaining available for evaluation, 15 converted to positive tuberculin status, of which two developed active disease.

The overall conversion rate for these physicians was 6.79%. In addition, the authors observed a positive correlation between the rate of conversion and the duration of postgraduate training. The conversion rate increased with the duration of training, beginning with a cumulative percentage of conversion of 2.06% in the first year, 8.62% in the 2nd year, 11.11% in the third year and 14.29% in the fourth year, resulting in a linear conversion rate of 3.96% per year. As noted by the authors, this linear increase suggests the hospital environment as the source of the infection. In addition, the prevalence rate of conversions in the hospital (17.24%) was much higher than would have been expected in the community for individuals of the same age.

The authors suggested that these high rates of conversion may have been a result of the fact that the hospital in this study encounters 5 to 10 times more active TB cases than most other urban hospitals. In addition, the physicians in training also are expected to be the first in line to perform physical evaluations and evaluate body fluids and secretions. While the authors did not go into detail about what, if any, TB infection control precautions were taken by these physicians in training, they did note that the evaluation of body fluids and secretions was often done in poorly ventilated and ill-equipped laboratories.

Increased rates of conversion were observed among employees in a New Orleans hospital in a 1986 study by Ktsanes et al. (Ex. 7-6). Similar to Miami, New Orleans also has a high rate of TB in the community. This study examined the skin test conversions among a cohort of 550 new employees who were followed for five years after assignment to the adult inpatient services. Of these 550 employees who were initially tuberculin negative, 17 converted to positive status over the five-year study period, resulting in an overall five-year cumulative conversion probability of 5.2%.

Regression analyses were done to examine potential contributing factors. Factors examined in the regression model included race, job, age at employment, and department. Only race (i.e., black vs. white employees) and job (i.e., nursing vs. other jobs) were found to be associated with skin test conversion. To further examine the potential job effect, conversions among blacks in nursing and blacks in other jobs were compared. Overall, the cumulative probability of converting was higher among blacks in nursing, suggesting that the acquired infections resulted from employment at the hospital rather than from the community at large. The authors thus concluded that there is an increased risk of occupational transmission of TB in TB-prevalent areas for those in close patient contact jobs.

In 1989, Haley *et al.* (Ex. 5-16) conducted a case study of a TB outbreak among emergency room personnel at a Texas hospital. In this study, a 70 year old male diagnosed with pulmonary TB and undergoing treatment was diverted, due to respiratory arrest, to Parkland Memorial Hospital while in route to another hospital. The man was admitted to the emergency room for approximately 4 hours until he was stabilized. Afterwards, the patient was placed in an intensive care unit, where he remained for 2 months until his death.

Six cases of active TB developed among emergency room employees after exposure to the TB patient, i.e., the index case. Five of these were among nurses who recalled contact with the index patient and a sixth case was an orderly who may have been infected from one of the employee TB cases. In addition, a physician exposed while administering treatment in the intensive care unit also developed active disease.

Skin test conversions were evaluated for the 153 employees of the emergency room. Of 112 previously negative employees, 16 had positive skin tests, including 5 nurses diagnosed with active TB. Fifteen of the conversions were a result of exposure to the index case. Skin tests were also evaluated for physicians in the intensive care unit. Of 21 resident physicians, two of whom had intubated the index patient, five had newly positive reactions to the tuberculin skin tests. One of the remaining three residents later developed active disease.

The authors attributed the outbreak to several factors. First, the index case had a severe case of pulmonary TB in which he produced copious amounts of sputum. Second, sixty percent of the emergency room air was recirculated without filtration adequate to remove TB bacilli, allowing for the recirculation of contaminated air. Finally, employees in the emergency room were provided surgical masks that were ineffective for protecting against transmission of airborne TB droplet nuclei. This study illustrates that the lack of effective measures for controlling TB transmission can result in the infection and development of active disease in a relatively high number of employees even after exposure to only one case of active TB.

Similarly, the lack of effective controls while performing high-hazard, cough-inducing procedures on individuals with infectious TB has also been shown to result in an increased risk of TB transmission. A 1990 report by Malasky *et al.* (Ex. 7-41) investigated the potential for TB transmission from high-hazard procedures by examining tuberculin skin test conversion rates among pulmonary physicians in training. In this study, questionnaires were sent annually, for 3 years, to training programs located in the top 25 cities for TB in 1983. The purpose of the study was to compare the conversion rates of pulmonary disease fellows to the conversion rates of infectious disease fellows. It was presumed that both groups have contact with patients with TB but that pulmonary disease fellows are usually more involved with invasive procedures such as bronchoscopies. Information requested on the questionnaires included the type of fellowship (i.e., pulmonary or infectious disease fellow), prior tuberculin skin test status, tuberculin status by the Mantoux technique at the end of the 3 year fellowship program, history of BCG vaccination, age, sex and ethnicity. In addition, the pulmonary disease fellows were asked to give information on the number of bronchoscopies they performed and their use of masks during the procedure.

Fourteen programs submitted data that were usable. Only programs that had both pulmonary and infectious disease fellows in the same system were used for the study. From this information, it was observed that 7 of 62 (11%) of the pulmonary fellows at risk converted their tuberculin skin test from negative to positive during the two year training period. In contrast, only 1 of 42 (2.4%) of the infectious disease fellows converted. The expected conversion rate from previous surveys was 2.3%. In addition, the pulmonary disease fellows were grouped according to tuberculin skin status. Skin test status was evaluated for its relationship to the number of bronchoscopies performed and the pattern of mask usage. No correlations were found with these factors and tuberculin skin status at the end of the fellowship. The authors suggested that the lack of correlation between mask usage during bronchoscopies and skin test conversion implies that masks worn by physicians may be inadequate. While little information was presented to evaluate this suggestion, the study does suggest that high-hazard procedures such as bronchoscopies that induce coughing, performed under uncontrolled conditions, present a risk for TB transmission.

Pearson *et al.* (1992) conducted a case-control study to investigate the factors associated with the development of MDR-TB among patients at a New York City hospital (Ex. 5-24). As a part of this study, tuberculin skin test conversion rates were compared among health care workers assigned to wards where patients with TB were frequently admitted (e.g., HIV unit, general medical ward, respiratory therapy) or rarely admitted (operating room, orthopedic ward, outpatient clinic, psychiatry ward). In addition, infection control procedures and ventilation systems were evaluated.

Of 79 health care workers who were previously negative, 12 (15%) had newly positive skin tests. Those health care workers who were assigned to wards where patients with TB were frequently admitted were more likely to have skin test conversions (i.e., 11 of 32) than health care workers assigned to wards where patients with TB were rarely admitted (i.e., 1 of 47).

Evaluations of the infection control procedures and ventilation systems revealed that patients who were receiving isolation precautions for suspected or confirmed TB were allowed to go to common areas if they wore a surgical mask. However, many of the patients did not keep their masks on when out of their rooms. In addition, neither the isolation rooms nor rooms used for cough-inducing procedures were under negative pressure, thus allowing contaminated air to exhaust to the adjacent corridors.

Edlin *et al.* (1992) (Ex. 5-9) investigated an outbreak of MDR-TB in a New York hospital among patients with acquired immunodeficiency syndrome (AIDS). This study compared the exposure period of AIDS patients diagnosed with MDR-TB to the exposure period of AIDS patients with drug-susceptible TB. The date of diagnosis was defined as the date the sputum sample was collected from which tuberculosis bacteria were grown in culture. Patients were assumed to be infectious two weeks before and two weeks after the date of diagnosis. The period of exposure was the period in which the patient may have been infected with TB. Because of the rapid progression from infection to disease, the exposure period was defined as 6 months preceding the date of diagnosis, excluding the last two weeks.

The patients with MDR-TB were found to be more likely to have been hospitalized during their exposure periods. Those who were hospitalized were more likely to have been on the same ward and on the same day as a patient with infectious TB and were more likely to have been near a room housing an infectious patient. Examination of the infectious patients' rooms revealed that only 1 of 16 rooms had negative pressure. Based on this evidence, the authors concluded that the observed cases of MDR-TB were a likely result of infections acquired in the hospital (i.e., primary TB) rather than as a result of the reactivation of infections acquired in the past. The authors attributed these nosocomial infections to the lack of adherence to recommended infection control procedures.

While the primary focus of this study was to investigate the transmission of TB among patients, the increased likelihood of nosocomial infections among patients in the hospital would seem equally likely to apply to health care workers working in the same environment. A survey of tuberculin skin test conversions revealed an 18% conversion rate for health care workers who previously had negative skin tests and were present during this outbreak of MDR-TB. Although no statistics were reported, the authors stated that the pattern of skin test conversions suggested an ongoing risk over time rather than a recent increase during the outbreak period.

Based on an earlier 1990 report from the CDC (Ex. 5-22), Beck-Sague *et al.* 1992 (Ex. 5-21) conducted a case-control study to investigate an outbreak of MDR-TB among the staff and patients in a HIV ward and clinic of a Miami hospital. As part of the overall study the authors compared the skin test conversion rates of health care workers in the HIV ward and clinic to the skin test conversion rates of health care workers in the thoracic surgery ward where TB patients were rarely seen. In addition, the authors also evaluated the relationship between the presence of patients with infectious MDR-TB and patients with infectious drug-susceptible TB on the HIV ward and the risk of skin test conversion among the HIV ward health care workers. Infection control procedures in the HIV ward and clinic were also examined.

All patients with suspected or confirmed TB were placed in isolation. However, some patients whose complaints were not primarily pulmonary and whose chest X-rays were not highly suggestive of TB were not initially suspected of TB and were not placed in isolation. Patients who were admitted to isolation rooms were allowed to leave TB isolation 7 days after the initiation of chemotherapy regardless of clinical or bacteriologic response. Thus, in some instances, patients with MDR-TB were allowed to leave isolation while they were still infectious, before drug resistance was recognized. In addition, patients in isolation rooms sometimes left the doors open, left their rooms, and/or removed their masks while outside their rooms. Patients with TB who were readmitted to the HIV ward and who were receiving anti-TB drugs were not admitted to isolation. In some cases, these patients were later found to have infectious MDR-TB.

An environmental assessment of the ventilation revealed that among 23 rooms tested with smoke tubes, 6 had positive pressure and many of the rooms under negative pressure varied from negative to positive depending on the fan setting and whether the bathroom door was open. Aerosolized pentamidine administration rooms were also found to have positive pressure relative to adjacent treatment areas. In addition, the sputum induction rooms were found to recirculate air back to the HIV clinic.

Skin test conversions were evaluated for all health care workers (i.e., nurses and clerical staff) who tested negative on the tuberculin skin test before the outbreak period, March 1988 through April 1990. Health care workers on the HIV ward and in the HIV clinic exhibited a significantly higher rate of skin test conversion than health care workers on the thoracic surgery ward (e.g., 13/39 vs. 0/15). Ten of the conversions occurred among the 28 health care workers in the HIV ward. Among these health care workers, the authors reported a significant correlation between the risk of infection in health care workers and the number of days that patients with infectious MDR-TB were hospitalized on the HIV ward. No correlation was observed between the risk of infection among health care workers on the HIV ward and the number of days that patients with infectious drug-susceptible TB were hospitalized on the ward.

Based on skin test conversions and the evaluation of infection control practices in the HIV ward and clinic, the authors concluded that the health care workers most likely were infected by patients on the HIV ward with MDR-TB. The factors most likely contributing to this increased risk of infection included: (1) The prolonged infectiousness and greater number of days that patients with infectious MDR-TB were hospitalized, (2) the delayed recognition of TB and failure to suspect infectious TB in patients receiving what proved to be ineffective anti-TB treatment, (3) the inadequate duration of, and lapses in, isolation precautions on the HIV ward, and (4) the lack of negative pressure ventilation in isolation and treatment rooms. While the evidence in this study primarily points to the transmission of MDR-TB from patients to health care workers, many of the problems identified with infection control procedures and ventilation would also increase the risk of acquiring drug-susceptible TB.

In addition to MDR-TB outbreak investigations in Miami, in 1993 the CDC reported an outbreak in New York City in which health care workers became infected after being exposed to patients with MDR-TB (Ex. 6-18). In this investigation, for the period December 1990 through March 1992, 32 patients were identified with MDR-TB. Twenty-eight of these patients had documented exposure to an undiagnosed infectious MDR-TB patient while all of them were in the HIV ward of the hospital.

During November 1991, health care workers who were assigned to the HIV inpatient unit and who were also previously negative on the tuberculin skin test, were given an additional skin test. Of 21 health care workers tested, 12 (57%) had converted to positive status (7 nurses, 4 aides and 1 clerical worker). None of the health care workers had used respiratory protection.

An investigation of infection control practices revealed that of 32 patients with MDR-TB, 16 were not initially suspected of TB and in these cases isolation precautions either were not used or were instituted late during the patients' hospitalization. In addition, patients who were admitted to isolation frequently left their rooms and when in their room the doors were frequently left open. Moreover, all rooms were found to be under positive pressure relative to the hall. Thus, similar to the findings in Miami, the results of this study indicate that the inability to properly isolate individuals with MDR-TB and also the use of inadequate respiratory protection may increase the risk of infection among health care workers.

Undiagnosed cases may also present a significant source for occupational transmission of TB. A case study by Cantanzaro (Ex. 5-14) described an outbreak of TB infection among hospital staff at a San Diego hospital where the hospital staff were exposed to a single patient with undiagnosed TB. In this case, a 64 year old man suffering from generalized seizures was transferred from a local jail to the emergency room and later admitted to a four bed intermediate care unit. While in the intermediate care unit he was treated with anticonvulsants but continued to have seizures accompanied with vomiting. He was therefore placed in intensive care where he underwent a variety of procedures including bronchoscopies and endotracheal intubation. During his stay, he received frequent chest therapy and suctioning. Three sputum samples were taken from the patient for smears and cultures. All AFB smears were negative. However, two cultures were positive for tuberculosis.

Despite the presence of positive cultures the patient was not diagnosed with active TB. The problem was not recognized until a physician on staff later developed symptoms of malaise and slight cough and requested a tuberculin skin test and was found to be positive. Because the physician had been tuberculin negative 8 months earlier, a contact investigation was initiated. As a part of this investigation, all employees who previously had negative tuberculin tests and who also worked in the intermediate and intensive care units where the patient had been treated were given repeat skin tests. Of 45 employees who previously had negative tuberculin skin tests, 14 (31%) converted to positive status (6 physicians, 3 nurses, 2 respiratory therapists and 1 clerk). Ten of these conversions were among the 13 previously tuberculin negative staff members who were present at the time bronchoscopies were conducted (10/13=76.9%). Four of the conversions were among 32 susceptible staff members who were not present at the bronchoscopies (4/32=12.5%). The author thus concluded that being present during the bronchoscopy of the patient was a major risk factor in acquiring the TB infection. However, the evidence did not show a significant correlation between skin test conversion and the type of exposure, i.e., close (administered direct contact) versus casual (in the room) contact. Thus, people who were present in the room during the bronchoscopy had an equal risk of infection as those administering direct patient care, presumably, as the author suggests, because droplet nuclei can disperse rapidly throughout the air of a room.

Similarly, Kantor et al. (Ex. 5-18) reported an outbreak of TB infection among hospital staff exposed to a single undiagnosed case of TB. The index case in this investigation was a 50 year old man who was admitted for lung cancer and was receiving chemotherapy, steroids and radiation treatment. After a month of treatment, the patient complained of a cough and chest pain and was found to have emphysema requiring additional drug treatment and a chest tube. However, even after the emphysema resolved, the patient complained of weakness, loss of appetite and fever. A sputum culture and smear were conducted for mycobacteria and found to be negative. Lung X-rays were found to be irregular but were attributed to the lung cancer. Upon his death the autopsy revealed extensive necrosis in the lung but tuberculosis was not suspected. Thus, no cultures for mycobacteria were performed and no infection control procedures were initiated. It was only upon histological examination of tissue samples one month later that the presence of TB was confirmed. Five months later one of the staff performing the autopsy developed active TB. His only history of exposure was to the index case.

As a result, a contact investigation was initiated for hospital personnel who had shared air with the patient during his stay, including the autopsy staff. Of susceptible hospital staff (i.e., those not previously found to react positive to the tuberculin skin test), infection developed in 9 of 56 (16%) exposed employees (4 autopsy staff, 4 nursing staff and 1 radiology staff). Only 3 of 333 unexposed personnel were found to have converted to positive tuberculin status at the hospital during the same period of investigation, thus indicating a 17.8 fold increase in the infection rate for the exposed group.

Undiagnosed cases of TB at time of autopsy were also indicated as the likely cause for development of active TB among staff and students in an autopsy room in a Swedish hospital (Ex. 5-19). In this study, three medical students and one autopsy technician, who were present during the autopsy of a patient with previously undiagnosed pulmonary TB, developed active TB. Both the medical students and the autopsy technician had previously received the BCG vaccine but none had any other known contact with a tuberculosis subject. Thus, it was concluded that the tuberculosis infections were most likely to have been transmitted during the autopsy. The findings of this study further illustrate the risks that undiagnosed cases of active TB present to health care workers. The lack of recognition of an active case of TB often results in a failure to initiate appropriate infection control procedures and provide appropriate personal protective equipment. In addition, this study illustrates that, while TB is most often transmitted by individuals with infectious pulmonary TB who generate droplet nuclei when they cough or speak, the autopsy procedures on deceased individuals with pulmonary TB may also aerosolize bacteria in the lungs and generate droplet nuclei.

Exposure during autopsy procedures was also suspected as a possible route of TB transmission in an upstate New York Medical Examiner's Office (Ex. 7-152). This Medical Examiner's Office conducted autopsies on deceased inmates from upstate New York prisons. In 1991, the same year that an outbreak of MDR-TB occurred among inmates from an upstate New York prison, the Medical Examiner's office conducted autopsies on 8 inmates with TB, six of whom had infectious MDR-TB at death and who were also HIV positive and had disseminated TB disease.

Skin tests were administered to employees who had worked for at least one month during 1991 at the Medical Examiner's Office. Among 15 employees who had originally tested negative on a baseline skin test, 2 were found to have converted. These two employees worked as morgue assistants and had recent documented exposure to persons with extensive disseminated MDR-TB. No potential exposure to TB outside the Medical Examiner's Office could be found.

The autopsy area of the office had a separate ventilation system. However, air was returned to a common air plenum, allowing the air to mix between the autopsy area and other areas of the office. In addition, the autopsy room was found to be at positive pressure relative to the adjacent hallway. Employees performing or assisting at autopsies on persons known to be infected with HIV were required to wear plastic gowns, latex gloves and surgical masks. Particulate respirators were not required until November of 1991, after the installation of germicidal UV lamps. However, this was after the last MDR-TB autopsy. This study suggests that the conversion of these two morgue assistants occurred as a result of exposure to aerosolized *M. tuberculosis* resulting from autopsy procedures, either as a result of participation in an autopsy in the autopsy area or from exposure to air contaminated with aerosolized *M. tuberculosis* that was exhausted into other areas of the Medical Examiner's Office.

In addition to autopsy procedures, other procedures, such as the irrigation of abscesses at sites of extrapulmonary TB, can result in the generation of droplet nuclei. An outbreak investigation in an Arkansas hospital (Ex. 5-17) reported the transmission of TB among hospital employees exposed to a patient with a tuberculous abscess of the hip and thigh. In this study, the source case was a 67 year old man who was admitted to the hospital with a fever of unknown origin and progressive hip pain. The patient did not present any signs of pulmonary TB; however, the examination of soft tissue swelling in the hip area revealed an abscess that required drainage and irrigation. Due to the suspicion of TB, specimens for AFB smear and culture were obtained and the patient was placed in isolation. While in isolation, drainage from the abscess continued and irrigation of the abscess cavity was initiated on an 8-hour schedule. After four days, acid fast bacilli were observed in the AFB smears and TB therapy was begun. The patient remained in isolation until his death except for three days that he spent in the Intensive Care Unit (ICU) due to high fever.

An investigation of skin test surveys among the hospital employees revealed 55 skin test conversions among 442 previously nonreactive employees and 5 conversions among 50 medical students. In addition, 5 of the employees who had conversions also had active TB, including one who developed a tuberculous finger lesion at the site of a needle-stick injury incurred during the incision and drainage of the patient's abscess. All the skin test converters, except for two, recalled exposure to the source case. Of the 442 susceptible employees, 108 worked at least one day on one of the floors where the patient stayed (i.e., the surgical ward, the medical floor of the patient's room and the ICU). Four (80%) of 5 surgical suite employees who had direct contact with the patient through their assistance with the incision and irrigation of the patient's abscess had skin test conversions. In addition, 28 (85%) of 33 employees on the general medical floor and 6 (30%) of 20 ICU employees had skin test conversions. All those employees converting recalled exposure to the patient, some of whom had no direct contact with the patient.

Environmental studies revealed that two of the areas in which the patient stayed during his hospitalization did not have negative pressure. The isolation room was under positive pressure relative to adjacent rooms and the corridor. In addition, the patient's cubicle in the ICU had neutral pressure relative to the rest of the ICU. Employees in these two areas had skin test conversions even in cases where there was no direct patient contact. The lack of negative pressure was thought to have significantly contributed to the dispersion of droplet nuclei generated from the irrigation of the tuberculous abscess. In the surgical ward, air was directly exhausted to the outside. However, all employees present in the surgical ward when the patient was being treated had direct contact with the patient. There was no indication that the surgical staff had taken any special infection control precautions or had worn any personal protective equipment.

Thus, similar to other outbreak investigations, the lack of appropriate ventilation and respiratory protection stand out as the key factors in the transmission of TB to employees who are exposed to individuals with infectious TB. Moreover, this particular case study demonstrates that certain forms of extrapulmonary TB in conjunction with aerosolizing procedures, e.g., the irrigation of a tuberculous abscess, have the potential for presenting significant airborne exposures to *M. tuberculosis*.

Other aerosolizing procedures have also shown evidence of presenting airborne exposures to *M. tuberculosis*. For example, tissue processing was associated with the skin conversion of two pathologists working at a community hospital in California (Ex. 6-27). In this case study, after autopsy, a 62 year old man who had died from bronchogenic carcinoma was discovered to have a caseating lung lesion. A stain revealed a heavy concentration of acid-fast bacilli, which were identified in culture as *M. tuberculosis*. As a result, a contact investigation was initiated.

This investigation found twenty employees who had contact with the patient, including two pathologists and a laboratory assistant. All were given a tuberculin skin test and found to be negative. However, after follow-up skin testing three months later, the two pathologists had converted. Other than contact with the source case, the two had no other obvious sources of infection. One of the pathologists had been present at the autopsy. Both pathologists were present when the frozen lung sections were prepared. During this process, the lung tissue was sprayed with a compressed gas coolant, which created a heavy aerosol. Masks were not routinely worn during this tissue processing. The investigators suspected that this aerosol promoted the transmission of TB and was the likely cause of the observed infections.

While much of the health effects literature has focused on outbreaks of TB or MDR-TB, a more recent study investigated the status of infection control programs among "non-outbreak" hospitals (Ex. 7-147). Investigators from the Society of Health care Epidemiology of America (SHEA) and the CDC surveyed members of SHEA to assess compliance in the respondents' hospitals with the 1990 CDC Guidelines for Preventing the Transmission of TB in Health Care Facilities for the years 1989 to 1992. The survey included questions on tuberculin skin testing programs (e.g., frequency of testing, positivity at hire, and percent newly converted), AFB isolation capabilities (e.g., negative pressure, air changes per hour, HEPA filtration) and respiratory protection.

The survey showed that of the 210 hospitals represented by the SHEA members' survey results, 193 (98%) admitted TB patients from 1989 to 1992, 40% of which had one or more patients with MDR-TB. In addition, the proportion of hospitals caring for drug susceptible TB patients rose from 88% to 92% and the proportion of hospitals caring for MDR-TB patients rose from 5% to 30%. While the number of hospitals caring for TB patients increased, the majority of those hospitals cared for a small number of patients. In 1992, approximately 89% of the hospitals reported 0 to 25 patients per year, while approximately 5% reported greater than 100 patients per year.

Few hospitals reported routine tuberculin skin testing for each of the years surveyed. For example, while 109 (52%) of the responding hospitals reported tuberculin skin test results for at least one of the years from 1989 to 1992, only 63 (30%) reported results for each of these years. When examining the conversion rates over time from 1989 to 1992, the investigators limited their analysis to the 63 hospitals reporting skin test data for each of these 4 years. Among these hospitals the median percentage of employees newly converting to positive skin test status remained constant over the 4 year period at approximately 0.34% per year (i.e., 3/1000 per year). However, when including all hospitals in the analysis, from 1989 to 1992, the number of hospitals reporting conversion rates increased from 63 to 109 and the conversion rates increased from 0.26% (i.e., 2/1000) to 0.50% (i.e., 5/1000).

With regard to AFB isolation capabilities, 62% of 181 responding hospitals reported that they had isolation facilities consistent with the 1990 CDC TB Guidelines (i.e., single-patient room, negative pressure, air directly exhausted outside, and ≥6 air changes per hour). Sixty-eight percent of the reporting hospitals had isolation facilities meeting the first three of these recommendations. For respiratory protection, the majority of health care workers in the hospitals used surgical masks. However, there was an increase in the use of dust-mist or dust-mist-fume respirators. The use of dust-mist respirators increased from 1 to 13% from 1989 to 1992 and the use of dust-mist-fume respirators increased from 0 to 10% for the same period. The only use of high efficiency particulate air (HEPA) filter respirators was by bronchoscopists and respiratory therapists at 4 hospitals.

As a second phase of this investigation, the survey responses were analyzed to determine the efficacy of the TB infection control programs among the member hospitals participating in the survey (Ex. 7-148). In this analysis, the reported conversion rates were compared to reported infection control measures (i.e., AFB isolation capabilities and respiratory protection). For purposes of comparison, hospitals were categorized as having either less than or ≥6 TB patients, less than or ≥437 beds, and admitting or not admitting MDR-TB patients.

Conversion rates were higher among health care workers from hospitals with ≥437 beds than among health care workers from smaller hospitals (0.9% vs. 0.6%, p≤0.05). This difference was more pronounced among "higher-risk" health care workers (i.e., health care workers including bronchoscopists and respiratory therapists). "Higher-risk" health care workers from hospitals with 437 or more beds had a 1.9% conversion rate compared to a conversion rate of 0.2% for "higher-risk" health care workers from smaller hospitals. Similarly, health care workers from hospitals where 6 or more TB patients were admitted per year had higher conversion rates than health care workers from hospitals with fewer than 6 TB patients per year (e.g., 1.2% vs. 0.6%).

For hospitals with 6 or more TB patients, conversion rates also varied depending on the level of TB infection control practices that were in place in the hospital. For example, among hospitals with 6 or more TB patients and whose AFB isolation capabilities included at least single-room occupancy, negative pressure and directly exhausted air, the conversion rates among health care workers were lower than the conversion rates among health care workers at hospitals with 6 or more TB patients but which did not have similar isolation capabilities (0.62% vs. 1.83%, p=0.03). For respiratory protection, however, no differences in conversion rates were observed among health care workers wearing surgical masks (0.94%) and health care workers using submicron surgical masks, dust-mist respirators or dust-mist-fume respirators (0.98%). Very few survey respondents reported use of HEPA filter respirators. For example, only four hospitals reported use of any HEPA respirators, and these were not the predominant type of respiratory protection used (Ex. 7-147). Thus, it is not possible to evaluate the efficacy of these particulate respirators in reducing conversion rates from the reported survey data.

For hospitals with fewer than 6 TB patients or with fewer than 437 beds, no differences in conversion rates were reported among health care workers from hospitals that had implemented AFB isolation capabilities such as single-room occupancy, negative pressure, or directly exhausted air and those hospitals that had not. The investigators suggested that this finding may support contentions that the efficacy of TB infection control measures vary depending on characteristics of the hospital or community exposure. However, given the small sample size of the survey, as well as the reduced potential for exposure in hospitals with fewer than 6 TB patients per year, it would be difficult to detect any differences in conversion rates among health care workers from hospitals with or without certain levels of infection control. Where more opportunity does exist for exposure (e.g., hospitals with ≥6 TB patients), this analysis does show that the implementation of TB infection control procedures can reduce the transmission of TB among health care workers.

HOSPITALS—SUMMARY

In summary, the evidence clearly shows that in hospital settings, employees are at risk of occupational exposure to TB. Various studies and TB outbreak investigations have shown that employees exposed to individuals with infectious TB have converted to positive tuberculin skin status and in some cases have developed active disease. In these reports, a primary factor in the transmission of TB has been a failure to promptly identify individuals with infectious TB so that appropriate infection control measures could be initiated to prevent employee exposure. In addition, another major factor identified as contributing to occupational exposures was the lack or ineffective implementation of appropriate exposure control methods (e.g., lack of negative pressure in isolation rooms, lack of appropriate respiratory protection for exposed employees, performance of high-hazard procedures under uncontrolled conditions). The lack of early identification and appropriate control measures resulted in the exposure and subsequent infection of various hospital employees. These employees included not only health care providers administering direct patient care to individuals with infectious TB, but also hospital staff providing support services to the infectious individuals, hospital staff working in adjacent areas of the hospital using shared air, autopsy staff and laboratory staff working with infected culture and tissue samples.

OTHER OCCUPATIONAL SETTINGS

While hospitals have been historically recognized as the primary type of work setting where TB presents an occupational hazard, there are other work settings where the transmission of TB presents a hazard to workers. There are a variety of occupational settings in which workers can reasonably be anticipated to encounter individuals with active TB as a part of their job duties. Several work settings have been identified by the CDC where exposure to TB presents an occupational hazard: correctional facilities, long-term care facilities for the elderly, homeless shelters, drug treatment centers, emergency medical services, home-health care, and hospices. Similar to the hospital setting, these work settings have a higher number of individuals with active TB than would be expected for the general population. Many of the clients of these work settings have many characteristics (e.g., high prevalence of TB infection, high prevalence of HIV infection, intravenous drug use) that place them at an increased risk of developing active TB. These types of work settings are also similar to hospitals in that workers at these sites may also provide medical services and perform similar types of high-hazard procedures that are typically done in a hospital setting.

In addition to employees who provide medical services in these other types of work settings, there are other types of workers (e.g., guards, admissions staff, legal counsel for prisoners) who may also be exposed to individuals with infectious TB. Similar to hospitals, these work settings have an over-representation of populations at high risk for developing active TB, e.g., individuals infected with HIV, intravenous drug users, elderly individuals, and individuals with poor nutritional status and who are medically underserved. In addition to having a higher percentage of individuals with TB infection and a higher percentage of individuals at an increased risk for developing active TB, many of these work settings also share environmental factors that facilitate the transmission of TB, such as overcrowding and inadequate ventilation, which increases the occupational hazard. The following discussion describes some of the studies available in the literature that have examined the occupational transmission of TB in other occupational settings such as those listed above. Not all the settings listed by the CDC as places where TB transmission may be likely to occur have been adequately studied and thus can not be included in this discussion. However, the discussion of the following sectors clearly demonstrates that the occupational transmission of TB is not limited to the hospital setting. Occupational settings where there is an increased likelihood of exposure to aerosolized *M. tuberculosis* present the same types of occupational hazards as have been documented in the hospital setting.

CORRECTIONAL FACILITIES

Many correctional facilities have a higher incidence of TB cases than occur in the general population. For example, the CDC reported that the incidence of TB among inmates of correctional facilities was more than three times higher than that for nonincarcerated adults aged 15-64, based on a survey of TB cases in 1984 and 1985 by 29 state health departments (Ex. 3-33). In particular, among inmates in the New York correctional system, the TB incidence increased from an annual average of 15.4 per 100,000 during 1976 to 1978 to 105.5 per 100,000 in 1986 (Ex. 7-80) to 156.2/100,000 for 1990-1991 (Ex. 7-137). Similarly, in 1987, the incidence of TB among inmates in New Jersey was 109.9 per 100,000 (approximately 11 times higher than the general population in New Jersey) and in California the incidence of TB among inmates was 80.3 per 100,000 (approximately 6 times higher than that for the general population for California) (Ex. 3-33). In 1989, the CDC reported that since 1985, eleven known outbreaks of TB have been recognized in prisons (Ex. 3-33).

The increased incidence of TB in correctional facilities has been attributed to several factors (Ex. 7-25). One, correctional facilities have a higher incidence of individuals who are at greater risk for developing active TB. For example, the population in prisons and jails may be dominated by persons from poor and minority groups, many of whom may be intravenous drug users. These particular groups may also suffer from poor nutritional status and poor health care, factors that place them at increased risk of developing active disease. Two, special types of correctional facilities, such as holding facilities associated with the Immigration and Naturalization Services, may have inmates/detainees from countries with a high incidence of TB. For foreign-born persons arriving in the U.S., the case rate of TB in 1989 was estimated to be 124 per 100,000, compared to an overall TB case rate of 9.5 per 100,000 for the U.S. (Ex. 6-26). In 1995, TB cases reported among the foreign born accounted for 35.7% of the total reported cases, marking a 63.3% increase since 1986 (Ex. 6-34). Three, many correctional facilities have a high proportion of individuals who are infected with HIV. The CDC reported that in addition to the growing increase in AIDS among prisoners, the incidence of AIDS in prisons is markedly higher than that for the U.S. general population. In 1988, the incidence of AIDS cases in the U.S. population was 13.7 per 100,000 compared to an estimated aggregate incidence for state/federal correctional systems of 75 cases per 100,000 (Ex. 3-33). Individuals who are infected with HIV or who have AIDS are at an increased risk of developing active TB due to their decreased immune capacity. The likelihood of pulmonary TB in individuals with HIV infection is reflected in the CDC's Revised Classification System for HIV infection (Ex. 6-30). In this revised classification system, the AIDS surveillance case definition was expanded to include pulmonary TB. Moreover, X-rays of individuals infected with HIV who have TB often exhibit radiographic irregularities that make the diagnosis of active TB difficult (Exs. 7-76, 7-77, 7-78, and 7-79). HIV-infected individuals may have concurrent pulmonary infections that confound the radiographic diagnosis of pulmonary TB. In addition, it may be difficult to distinguish symptoms of TB from *Pneumocystis carinii* pneumonia or other opportunistic infections. This difficulty in TB diagnosis can result in true cases of active TB going undiagnosed in this population. Undiagnosed TB has been shown to be an important cause of death in some patients with HIV infection (Ex. 7-76). Fourth, environmental conditions in correctional facilities can aid in the transmission of TB. For example, many prisons are old, have inadequate ventilation systems, and are overcrowded. In addition, inmates are frequently transferred both within and between facilities, thus increasing the potential for the spread of TB infection among inmates and staff. This increased potential for mobility among inmates also enhances the likelihood that inmates undergoing therapy for active disease will either discontinue their treatment or inadequately follow their prescribed regime of treatment. The inadequacy of their treatment may give rise not only to relapses to an infectious state of active disease, but also potentially give rise to strains of MDR-TB. These strains of TB have a higher incidence of fatal outcome and are generally characterized by prolonged periods of infectiousness during which the risk of infection to others is increased.

The high incidence of TB among the inmate population presents an occupational hazard to the staff in these types of facilities. Recent outbreak investigations by the CDC have documented the transmission of TB to exposed workers. In an investigation of a state correctional facility in New York for 1991 (Exs. 6-3 and 7-136), eleven persons with TB were identified (10 inmates and one correctional facility guard). Nine persons (8 inmates and the guard) had MDR-TB. All eight inmates were HIV positive. The guard was HIV negative; however, he was also immunocompromised as a result of treatment for laryngeal cancer. Seven of the inmates and the guard died from MDR-TB. The eighth inmate was still alive and receiving treatment for MDR-TB 2 years after being diagnosed as having the disease. DNA analysis identified the strains of tuberculosis bacteria from these individuals to be identical.

The investigation revealed that the source case was an inmate who had been transferred from another prison where he had been previously exposed to MDR-TB. He arrived at the prison with infectious TB but refused evaluation by the infirmary staff. This inmate was placed in the general prison population where he stayed for 6 months until he was admitted to the hospital where he later died. However, before his hospitalization, he exposed two inmates living in his cell block who later developed MDR-TB. These two inmates continued to work and live in the prison until shortly before their final hospitalization. The other inmates who subsequently developed MDR-TB had several potential routes of exposure: social contact in the prison yard, contact at work sites in the prison, and contact at the prison infirmary where they shared rooms with other inmates before diagnosis with TB.

The guard who developed MDR-TB had exposure to inmates while transporting them to and from the hospital. The primary exposure for this guard apparently occurred when he was detailed outside the inmates' room during their hospitalization for MDR-TB. The inmates were hospitalized in an isolation room with negative pressure. However, upon investigation it was discovered that the ventilation system for the room had not been working correctly and had allowed air to be exhausted to the hospital corridors and other patient rooms.

A contact investigation in the prison was conducted to identify other inmates who might have been exposed during this outbreak of MDR-TB. Of those inmates with previous negative tuberculin skin tests and without active disease (306), ninety-two (30%) had documented skin test conversions. There was no tuberculin skin test program for prison staff; therefore, conversions among prison employees could not be evaluated.

The primary factors identified as contributing to this outbreak were deficiencies in identifying TB among transferred inmates, laboratory delays, and lapses in isolating inmates with active TB within the facility. Inmates with symptoms of active disease were not sent for evaluation in some cases until they became so ill they could not care for themselves. Some of these inmates were placed in the infirmary with other inmates until their diagnosis with TB. On other occasions, drug susceptibility testing was not reported until after an inmate's death, which means that appropriate patient management was not initiated.

As a result of this outbreak, a retrospective epidemiological investigation was conducted to examine the potential extent and spread of MDR-TB throughout the New York State prison system during the years 1990-1991 (Ex. 7-137). This investigation revealed that 69 cases of TB were diagnosed in 1990 and another 102 were diagnosed in 1991, resulting in a combined incidence of 156.2 cases/100,000 inmate years for 1990 and 1991 combined. Of the cases, 39 were identified as being MDR-TB, 31 of which were shown to be epidemiologically linked. Thirty-three of the individuals with MDR-TB never received any treatment for MDR-TB, 3 were diagnosed at death, and 23 died before drug susceptibility results were known. These inmates were also discovered to be highly mobile. The 39 inmates lived in 23 different prisons while they were potentially infectious. Twenty transfers were documented for 12 inmates with potentially infectious MDR-TB (9 shortly before diagnosis, one after diagnosis with TB but before diagnosis with MDR-TB, and 2 after a diagnosis of MDR-TB).

Several factors were identified as contributing to the spread of MDR-TB throughout the New York prison system: delays in identifying and isolating inmates, frequent transfers without appropriate medical evaluation, lapses in treatment, and delays in diagnosis and susceptibility testing.

A similar investigation in a California state correctional institution identified three active cases of TB (two inmates and one employee) during September and October 1991 (Ex. 6-5). As a result, an investigation was commenced to determine whether transmission of TB was ongoing in the institution. Eighteen inmates with active TB were identified. TB in 10 of these inmates was recognized for the first time while they were in the institution during 1991, resulting in an annual incidence of TB of 184 per 100,000, a rate greater than 10 times that for the state (17.4 per 100,000). Two of the 10 inmates had negative tuberculin skin tests prior to their entry into the institution. Three of the cases were determined to have been infectious during 1991.

A review of skin test data revealed that for the 2944 inmates for whom skin test results were available, 324 tested positive for the first time while in the prison system. Of these, 106 were tuberculin negative before their entry into the prison system, 96 of which occurred in the previous two years.

The employee identified as having active TB had worked as a counselor on the prison's HIV ward, where he recalled exposure to one of the 3 infectious inmates. This employee could recall no known exposures outside the prison. Similarly, two other prison employees had documented skin test conversions while working at the prison. Neither recalled exposures outside the prison; one reported exposure to an inmate with possible TB.

No information was provided in this report as to whether any isolation precautions were implemented at this facility. However, the investigators concluded that their findings suggested the likelihood that transmission of TB had occurred in the prison. Their conclusion was based on the fact that a substantial number of skin test conversions were documented among the inmates and that at least two inmates with active TB became infected while at the prison.

The transmission of TB was also reported in another California prison among prison infirmary physicians and nurses and correctional officers (Ex. 6-6). In this investigation, an inmate with active MDR-TB spent 6 months during 1990-1991 in the infirmary. The infirmary had no isolation rooms and inmates' cells were found to be under positive pressure. Employees occasionally recalled wearing surgical masks when entering the rooms of TB patients.

An analysis of available skin testing data revealed that of the 21 infirmary health care providers, only 10 had been tested twice during the period from 1987 to 1990. Of these 10, two were newly positive, one of whom had recently converted in 1991 and had spent 5 months in the preceding year providing health care to the source case in this investigation. Another health care provider and a correctional officer who worked in the infirmary also were identified as having newly converted while at the prison. There was no yearly skin test screening, and thus their conversions could have occurred at any time between 1987 and 1991. However, 13 other inmates were diagnosed with pulmonary TB during that same period. An additional correctional officer who did not work in the infirmary also was found to have newly converted. His reported exposure occurred at a community hospital where he was assigned to an inmate with infectious TB. The officer was not provided with any respiratory protection. The lack of isolation precautions and the lack of appropriate respiratory protection suggest transmission of TB from infectious inmates in the infirmary to the prison staff, either as a result of exposure to the source case or other inmates with pulmonary TB who were also treated in the prison infirmary. Because of the lack of contact tracing or routine annual screening of inmates or staff, the full extent of transmission from the source case or other TB cases could not be determined.

Thus, similar to the evidence for the hospital setting, the evidence on correctional facilities shows that the failure to promptly identify individuals with infectious TB and provide appropriate infection control measures can result in the exposure and subsequent infection of employees with TB. These employees include the correctional facility infirmary staff, guards on duty at the facility, and guards assigned to escort inmates during transport to other facilities (e.g., outside health care facilities and other correctional facilities).

HOMELESS SHELTERS

Tuberculosis has also been recognized as a health hazard among homeless persons. The growth of the homeless population in the United States since the 1980s and the subsequent increase in the number of shelters for the homeless, furthers heightens the concern about the potential for the increased incidence and transmission of TB among the homeless, especially in crowded living conditions such as homeless shelters.

A number of factors are present in homeless shelters which increase the potential for the transmission of TB among the shelter residents and among the shelter staff. A high prevalence of TB infection and disease is common among many homeless shelters. This is not surprising, since the residents of these facilities usually come from lower socio-economic groups and often have characteristics that place them at high risk. Screening of selected clinics and shelters for the homeless has shown that the prevalence of TB infection ranges from 18 to 51% and the prevalence of clinically active disease ranges from 1.6 to 6.8% (Ex. 6-15). The CDC estimates this to be 150 to 300 times the nationwide prevalence rate (Ex. 6-17).

In addition to having a high prevalence of individuals with TB infection in the shelters, many of the shelter residents possess characteristics that impair their immunity and thus place them at a greater risk of developing active disease. For example, homeless persons generally suffer from poor nutrition, poor overall health status and poor access to health care. Many also suffer from alcoholism, drug abuse and psychological stress. Moreover, a significant portion of homeless shelter residents are infected with the HIV. In 1988, the Partnership of the Homeless Inc. conducted a survey of 45 of the nation's largest cities and estimated that there were between 5,000 and 8,000 homeless persons with AIDS in New York City and approximately 20,000 nationwide (Ex. 7-55). Due to these factors, homeless shelter residents are at increased risk of developing active disease. Thus, there is the increased likelihood that these individuals will be infectious as a result of active disease and thereby present a source of exposure for other homeless persons and for shelter employees.

In addition to having factors which increase their risk of developing active TB disease, homeless persons also are a very transient population. Because they are transient, homeless persons are more likely to discontinue or to erratically adhere to the prescribed TB therapy. Inadequately adhering to TB therapy can result in relapses to an infectious state of the disease or the development of MDR-TB. Both outcomes result in periods of infectiousness, during which they present a source of exposure to other residents and staff. In addition, environmental factors at homeless shelters, such as crowded living conditions and poor ventilation, facilitate the transmission of TB.

Outbreaks of TB among homeless shelter residents have been reported. For example, during 1990, 17 individuals with active pulmonary TB were identified among residents of homeless shelters in three Ohio cities: Cincinnati, Columbus, and Toledo (Ex. 7-51). In Cincinnati, 11 individuals with active TB were identified in a shelter for homeless adults. The index case was a man who had resided at the shelter and later died from respiratory failure. He was not diagnosed with TB until his autopsy. Of these 11 individuals, of which the index case was one, 7 were determined to be infectious. There was no indication as to whether any infection control measures were in place in the shelter. DNA analysis of 10 individual *M. tuberculosis* isolates showed identical patterns. The similarity among these DNA patterns suggested that transmission of the TB occurred in the shelter.

While the primary focus of this investigation was on the active cases reported among the residents in this Cincinnati shelter, the risk of transmission identified in this shelter also would apply to the shelter staff. Possible transmission of TB infection from the infectious individuals to the shelter staff might have been identified through tuberculin skin test conversions. However, no tuberculin skin test information for the staff was reported in this investigation.

Tuberculin skin testing results were reported in the investigation of a Columbus, Ohio shelter. In this investigation, a resident of a Columbus homeless shelter was identified with infectious pulmonary TB at the local hospital in March of 1990. The patient also had resided in a shelter in Toledo. As a result, a city-wide TB screening was initiated from April to May 1990 among the residents and staff of the city's men's shelters. Tuberculin skin tests were conducted on 363 shelter residents and 123 shelter employees. Among 81 skin-tested residents of the shelter in which the index case had resided, 32 (40%) were positive compared to 47 (22%) of 210 skin-tested residents of other shelters in Columbus who had positive skin test reactions. Similarly, among 27 employees of the shelter where the index case resided, 7 (26%)

had positive skin test reactions compared to 9 (11%) of 85 employees in other men's shelters. These skin test results suggest an increased risk of transmission of TB among residents and employees of the homeless shelter where the index case resided. However, due to the lack of baseline skin test information among these residents and employees it is not possible to determine when their conversion to positive status occurred and whether this index case was their source of exposure. These results, however, do indicate a high prevalence of TB infection among homeless residents (e.g., 40% and 22%). Many of these individuals are likely to have an increased risk of developing active TB and, as a result, they may present a source of exposure to residents and staff.

The transmission of TB has also been observed among residents and staff of several Boston homeless shelters (Exs. 7-75 and 6-25). From February 1984 through March 1985, 26 cases of TB were confirmed among homeless residents of three large shelters in Boston. Nineteen of the 26 cases occurred in 1984, thus giving an incidence of approximately 317 per 100,000, 6 times the homeless case rate of 50 per 100,000 reported for 1983 and nearly 16 times the 1984 case rate of 19 per 100,000 for the rest of Boston (Ex. 6-25).

Of the 26 cases of TB reported, 15 had MDR-TB. Phage typing of isolates from 13 of the individuals with drug-resistant TB showed identical phage types, thus suggesting a common source of exposure. As a result of this outbreak, a screening program was implemented in November 1984 over a four-night period. Of 362 people who received skin tests, 187 returned for reading, 42 (22%) were found to be positive and 3 were recent converters. Screening also was reported for the shelter staff at the three homeless facilities. At the largest of the three shelters, 17 of 85 (20%) staff members had skin test conversions. In the other two shelters, 3 of 15 (20%) and 3 of 18 (16%) staff members had skin test conversions.

Whereas MDR-TB was primarily involved in the outbreak in Boston, an outbreak of drug-susceptible TB was reported in a homeless shelter in Seattle, Washington (Ex. 7-73). From December 1986 to January 1987, seven cases of TB from homeless residents were reported to the Seattle Public Health Department. The report of 7 individuals with active TB in one month prompted an investigation, including: (1) A mass screening to detect undiagnosed cases, (2) phage typing of isolates from shelter clients to detect epidemiologically linked cases, and (3) a case-control study to investigate possible risk factors for the acquisition of TB.

A review of the case registries revealed that 9 individuals with active TB had been reported from the homeless shelter for the preceding year and four cases in the year previous to that. As a result of the mass screening in late January 1987, an additional 6 individuals with active TB were detected. Phage typing of 15 isolates from the shelter-associated cases revealed that 6 individuals with active TB diagnosed around the time of the outbreak were of the same phage type, suggesting that there was a predominant chain of infection, i.e., a single source of infection. However, there also were other phage types, suggesting several sources of infection. Therefore, the investigators suggested that there was probably a mixture of primary and reactivated cases.

In addition to the similarity of phage types among TB cases, tuberculin skin testing results suggested the ongoing transmission of TB in the shelter. For example,

10 shelter clients who were previously tuberculin negative in May 1985 were re-tested in January 1987 and 3 (30%) had converted. In addition, 43 clients who were negative in January 1987 were re-tested in June 1987 or February 1988 and 10 (23%) had converted. Factors identified as contributing to the outbreak were the increased number of men with undiagnosed infectious pulmonary TB, the close proximity of beds in the shelter, and a closed ventilation system that provided extensive recircu-lation of unfiltered air.

As a result of the outbreak, a control plan was implemented. This plan included repetitive mass screening, repetitive skin testing, directly observed therapy, preven-tive therapy and modification of the ventilation system to incorporate UV light disinfection in the ventilation duct work. After the control plan was in place, five additional individuals with active TB were observed over a 2-year follow-up period.

While the primary focus in this study was on clients of the shelter rather than the shelter staff, the risk factors present in the shelter before implementation of the control plan would have also increased the likelihood for transmission of TB to shelter employees from infectious clients.

Thus, similar to correctional facilities, homeless shelters have a number of risk factors that facilitate and promote the transmission of TB (e.g., high incidence of infected residents with an increased likelihood of developing active disease, crowded living conditions and poor ventilation). Also, similar to correctional facilities, the evidence in homeless shelters shows that the failure to promptly identify homeless residents with infectious TB and the lack of appropriate TB control measures (e.g., lack of isolation precautions or prompt transfer to facilities with adequate isolation precautions) resulted in the transmission of TB to shelter employees.

LONG-TERM CARE FACILITIES FOR THE ELDERLY

Long-term care facilities for the elderly also represent a high-risk population for the transmission of TB. TB disease in persons over the age of 65 constitutes a large proportion of TB in the United States. Many of these individuals were infected in the past, before the introduction of anti-TB drugs and TB control programs when the prevalence of TB disease was much greater among the general population, and have harbored latent infection over their lifetimes. However, with advancing age, these individuals' immune function starts to decline, placing them at increased risk of developing active TB disease. In addition, they may have underlying disease or overall poor health status. Moreover, residents are often clustered together and group activities are often encouraged. TB case rates are higher for this age group than for any other. For example, the CDC reports that in 1987, the 6,150 cases of TB disease reported for persons ≥65 years of age accounted for 27% of the U.S. TB morbidity although this group only represented 12% of the U.S. population (Ex. 6-14).

Because of the higher prevalence of TB cases among this age group, employees of facilities that provide long-term care for the elderly are at increased risk for the transmission of TB. More elderly persons live in nursing homes than in any other type of residential institution. The CDC's National Center for Health Statistics reports that elderly persons represent 88% of the nation's approximately 1.7 million nursing home residents. As noted by the CDC, the concentration of such high-risk individuals in long-term care facilities creates a high-risk situation for the transmission of TB (Ex. 6-14).

In addition to having a higher prevalence of active TB, the recognition of TB in elderly individuals may be difficult or delayed because of the atypical radiographic appearance that TB may have in elderly persons (Exs. 7-59, 7-81, 7-82, and 7-83). In this situation, individuals with active TB may go undiagnosed, providing a source of exposure to residents and staff.

While the increased incidence of TB cases among the elderly in long-term care facilities may be a result of the activation of latent TB infections, the transmission of TB infection to residents and staff from infectious cases in the facilities has been observed and reported in the scientific literature.

For example, Stead et al. (1985) examined the reactivity to the tuberculin skin test among nursing home residents in Arkansas (Ex. 7-59). This study involved a cross-sectional survey in which tuberculin skin tests were given to all current nursing home residents. In addition, all newly-admitted nursing home residents were skin tested. For the three year period evaluated, 25,637 residents of the 223 nursing homes in Arkansas were tested.

Of 12,196 residents who were tested within one month of entry, only 12 percent were tuberculin positive, including those for whom a booster effect was detected. However, among the 13,441 residents for whom the first test was delayed for more than a month, 20.8% were positive. In addition, the results of retesting 9,937 persons who were tuberculin negative showed an annual conversion rate of approximately 5% in nursing homes in which an infectious TB case had been recognized in the last three years. In nursing homes with no recognized cases, the authors reported an annual conversion rate of approximately 3.5%. The authors concluded that their data supported the contention that tuberculosis may be a rather common nosocomial infection in nursing homes and that new infections with tuberculosis is an important risk for nursing home residents and staff.

Brennen et al. (Ex. 5-12) described an outbreak of TB that occurred in a chronic care Veteran's Administration Medical Center in Pittsburgh. This investigation was initiated as a result of two skin test conversions identified through the employee testing program. One converter was a nurse working on ward 1B (a locked ward for neuropsychiatric patients) and the other was a physician working in an adjacent ward, 1U, who also had significant exposure to ward 1B. The source of infection in this investigation was traced to two patients who had resided on ward 1B and who had either a delayed or undiagnosed case of TB. The contact investigation revealed 8 additional conversions among patients, 4 in ward 1B and 4 in wards 2B and 4B (units on the floor above 1B).

Because the source cases were initially unidentified, no isolation precautions were taken. Smoke tracer studies revealed that air discharged from the window air conditioning unit of one of the source patients discharged directly into the courtyard. Air from this courtyard was the air intake source for window air conditioning units in the converters' room on ward 2B and thus was one of the possible sources of exposure.

In addition to the contact investigation on ward 1B and the adjacent units, hospital-wide skin testing results were evaluated. Of 395 employees tested, 110

(28%) were positive. The prevalence in the surrounding community was estimated to be 8.8%. Of those employees initially negative, 38 (12%) converted to positive status. Included among these were employees in nursing (18), medical (3), dental (1), maintenance/engineering (3), supply (1), dietary (9), and clerical (2) services.

Occupational transmission of TB was also reported in a nursing home in Oklahoma (Ex. 6-28). In August 1978, a 68 year old female residing in the east wing of the home was diagnosed with pulmonary TB. She was subsequently hospitalized. However, by that time she had already had frequent contact with other residents in the east wing. As a result, a contact investigation, in which all residents of the home were given skin tests, was initiated.

The investigation revealed that the reaction rate for residents in the east wing (34/48, 71%) was significantly higher than the reaction rates of residents living in the north and front wings (30/87, 34%). No baseline skin test information was presented for the residents to determine the level of conversion. However, it was noted that half of the nursing home residents were former residents of a state institution for the developmentally disabled. A 1970 tuberculin skin test survey of that institution had shown a low rate of positive reactions.

In addition to the nursing home residents, nursing home employees were also skin tested. Of the 91 employees tested, 61 (67%) were negative and 30 (33%) were positive. Similar to results observed among the residents, positive reaction rates were higher for employees who had ever worked in the east wing (50%) than for those who had never worked in the east wing (23%). Retesting of the employees 3 months later revealed 3 conversions. These results suggested that there may have been occupational transmission of TB in this facility.

Occupational transmission has also been observed in a retrospective study of residents and employees who lived or worked in an Arkansas nursing home between 1972 and 1981 (Ex. 7-83). In this retrospective study, investigators reviewed the skin testing and medical chart data collected over a 10-year period at an Arkansas nursing home. Among the nursing home residents who were admitted between 1972 and 1982, 32 of 226 residents (17%) who were initially tuberculin negative upon admittance became infected while in the home, based on conversion to positive after at least two previous negative tests. Twenty-four (63%) of these conversions were infected in 1975, following exposure to one infectious resident. This resident, who had negative skin tests on three previous occasions during his stay in the home, was not diagnosed with TB until after he was hospitalized because of fever, loss of weight and productive cough. The remaining 37% converted in the absence of a known infectious case. Thus, the authors suggested that nosocomial infections are likely to result from persons unsuspected of having TB.

Skin testing was also reviewed for employees of the nursing home. Questionnaires were completed by 108 full-time employees. Eleven of 68 employees with follow-up skin tests converted to positive skin status during the study period. Ten of the 11 (91%) converters reported that they had been in the nursing home in 1975, the same year in which many of the residents were also found to have converted from a single infectious case. In addition, employees working at least 10 years in the home had a higher percentage of conversions (9 of 22, 40%) than employees working less than 10 years (2 of 46, 4.4%). Based on the results of this study, the authors concluded that, in addition to occurrence of TB cases from the reactivation of latent infections among the elderly, TB can also be transmitted from one resident to another resident or staff. Consequently, TB must be considered as a potential nosocomial infection in nursing homes.

Thus, long-term care facilities for the elderly represent a high-risk situation for the transmission of TB. These types of facilities possess a number of characteristics that increase the likelihood that active disease may be present among the facility residents and may go undetected. Similar to other high-risk settings, the evidence shows that the primary factors in the transmission of TB among residents and staff have been the failure to promptly identify residents with infectious TB and initiate and adequately implement appropriate exposure control measures.

Drug Treatment Centers

Another occupational setting that has been identified as a high-risk environment for the transmission of TB is drug treatment centers. Similar to other high-risk sites, drug treatment centers have a higher prevalence of TB infection than the general population. For example, in 1989 the CDC funded 25 state and city health departments to support tuberculin testing and administration of preventive therapy in conjunction with HIV counseling and testing. In this project, 28,586 clients from 114 drug treatment centers were given tuberculin skin tests. Of those, 2,645 (9.7%) were positive (Ex. 6-8). When persons with previously documented positive tests were included, 4167 (13.3%) were positive.

There is also evidence to suggest that drug dependence is a risk factor for TB disease. For example, Reichman et al. (Ex. 7-85) evaluated the prevalence of TB disease among different drug-dependent populations in New York: (1) An in-hospital population, (2) a population in a local drug treatment center, and (3) a city-wide population in methadone clinics. For the in-hospital population of 1,283 patients discharged with drug dependence, 48 (3.74%) had active disease, for a prevalence rate of 3,740 per 100,000. In comparison, the TB prevalence rate for the total inpatient population was 584 per 100,000 and for New York City as a whole was 86.7 per 100,000. Screening of clients at a local drug treatment center in Harlem revealed a TB prevalence of 3750 per 100,000 in the drug-dependent population. Similarly, in the New York methadone program, the city-wide TB prevalence was 1,372 per 100,000. The authors also reported that although estimates of TB infection rates for both drug-dependent and non-drug dependent people were similar, the prevalence of TB disease among the drug-dependent was higher, thus suggesting that drug dependency may be a risk factor for disease.

Clients of drug treatment centers not only have a high prevalence of TB infection, a majority of them are intravenous drug users. Of the estimated 645,000 clients discharged each year from drug treatment centers, approximately 265,000 are intravenous drug users who either have or are at risk for HIV infection. In the Northeastern U.S., HIV seroprevalence rates of up to 49% have been reported (Ex. 6-8). These individuals are at increased risk of developing active TB disease.

To determine the risk of active TB associated with HIV infection, Selwyn et al. (Ex. 5-6) prospectively studied 520 intravenous drug users enrolled in a methadone maintenance program. In this study, 217 HIV seropositive and 303 seronegative intravenous drug users, who had complete medical records documenting their history of TB and skin test status, were followed from June 1985 to January 1988. On admission to the methadone program, and at yearly intervals, all patients were given tuberculin skin tests.

Forty-nine (23%) of the seropositive patients and 62 (20%) of the seronegative patients had positive reactions to the skin test before entry into the study. Among the patients who initially had negative skin tests, 15 of 131 (11%) seropositive patients and 62 of 303 (13%) seronegative patients converted to positive tuberculin status. While the prevalence and incidence rates of TB infection were similar for the two groups of patients, seropositive patients showed a higher incidence of developing active disease. Active TB developed in 8 of the seropositive subjects with TB infection (4%), whereas none of the seronegative patients with TB infection developed active TB during the study period.

Among individuals who are infected with HIV or who have AIDS, TB disease may be difficult to diagnosis because of the atypical radiographic appearance that TB may present in these individuals. In these individuals, TB may go undiagnosed and present an unsuspected source of exposure. Clients of drug treatment centers also may be more likely to discontinue or inadequately adhere to TB therapy regimens in instances where they develop active disease. As in other instances, this increases the likelihood of relapse to active disease or possibly the development of MDR-TB, both of which result in additional or even prolonged periods of infectiousness during which other clients or staff can be exposed.

There is evidence showing the transmission of TB in drug treatment facilities among both the clients and the staff. In a CDC case study (Ex. 5-6), a Michigan man who was living in a residential substance abuse treatment facility and was undergoing therapy for a previously diagnosed case of TB disease, was discovered by the local health department to have MDR-TB. As a result, a contact investigation was initiated at the drug treatment facility in which he resided.

Of the 160 clients and staff who were identified as potential contacts, 146 were tested and given tuberculin skin tests in November. No health screening program had been in place at the facility. The following March repeat skin tests were given. Of the 70 persons who were initially tuberculin negative and were still present in the facility, 15 (21%) had converted to positive status (14 clients and 1 staff member). The investigators noted that the number of converters may have been underestimated for two reasons. Many of the clients were at risk for HIV infection and thus may have been anergic and not responded to the tuberculin skin tests. In addition, nearly half of the clients who were initially negative were not available for repeat skin testing.

Several factors may have contributed to the observed conversions in this facility. For example, no health screening program was in place. Therefore, individuals with TB would go unidentified. In addition, the clients were housed in a building with crowded dormitories for sleeping. The only ventilation in this building was provided by opening windows and doors. Thus, environmental conditions were ideal for the transmission of TB.

Consequently, the high-risk characteristics of clients who frequent these centers (e.g., high prevalence of infection and factors increasing the likelihood of developing active disease) and environmental characteristics of the center (e.g., crowding and poor ventilation), lead to drug treatment centers being considered a high-risk setting for the transmission of TB. The available evidence shows that the failure to promptly identify clients with infectious TB and to initiate and properly implement exposure control methods (e.g., proper ventilation) resulted in the infection of clients and staff at these facilities.

CONCLUSION

The available evidence clearly demonstrates that the transmission of TB represents an occupational hazard in work settings where employees can reasonably be anticipated to have contact with individuals with infectious TB or air that may reasonably be anticipated to contain aerosolized *M. tuberculosis* as a part of their job duties. Epidemiological studies, case reports, and outbreak investigations have shown that in various work settings where there has been an increased likelihood of encountering individuals with active TB or where high-hazard procedures are performed, employees have become infected with TB and in some cases developed active disease. While some infections were a result of more direct and more prolonged exposures, other infections resulted from non-direct and brief or intermittent exposures. Because of the variability in the infectiousness of individuals with active TB, one exposure may be sufficient to initiate infection.

Several factors, common to many of these work settings, were identified as contributing to the transmission of TB: (1) Failure or delayed recognition of individuals with active TB within the facility, and (2) failure to initiate or adequately implement appropriate infection control measures (e.g., performance of high-hazard procedures under uncontrolled conditions, lack of negative pressure ventilation, recirculation of unfiltered air, and lack of appropriate respiratory protection). Thus, in work settings where employees can reasonably be anticipated to have contact with individuals with infectious TB or air that may contain aerosolized *M. tuberculosis* and where appropriate infection control programs are not in place, employees are at increased risk of becoming infected with TB.

Infection with TB is a material impairment of the worker's health. Even though not all infections progress to active disease, infection marks a significant change in an individual's health status. Once infected, the individual is infected for his or her entire life and carries a lifetime risk of developing active disease, a risk they would not have had they not been infected. In addition, many individuals with infection undergo preventive therapy to stop the progression of infection to active disease. Preventive therapy consists of very toxic drugs that can cause serious adverse health effects and, in some cases, may be fatal.

Although treatable, active disease is also a serious adverse health effect. Some TB cases, even though cured, may result in long-term damage to the organ that is infected. Individuals with active disease may need to be hospitalized while they are infectious and they must take toxic drugs to stop the progressive destruction of the infected tissue. These drugs, as noted above, are toxic and may have serious side effects. Moreover, even with advancements in treating TB, individuals still die from TB disease. This problem is compounded by the emergence of multidrug-resistant strains of TB. In these cases, due to the inability to find adequate drug regimens which can treat the disease, individuals remain infectious longer, allowing the disease to progress further and cause more progressive destruction of the infected tissue. This increases the likelihood of long-term damage and death.

PRELIMINARY RISK ASSESSMENT FOR OCCUPATIONAL EXPOSURE TO TUBERCULOSIS

INTRODUCTION

The United States Supreme Court, in the "benzene" decision (*Industrial Union Department, AFL-CIO* v. *American Petroleum Institute*, 448 U.S. 607 (1980)), has stated the OSH Act requires that, prior to the issuance of a new standard, a determination must be made, based on substantial evidence in the record considered as a whole, that there is a significant health risk under existing conditions and that issuance of a new standard will significantly reduce or eliminate that risk. The Court stated that

> "before he can promulgate any permanent health or safety standard, the Secretary is required to make a threshold finding that a place of employment is unsafe in the sense that significant risks are present and can be eliminated or lessened by a change in practices" (448 U.S. 642).

The Court in the Cotton Dust case (*American Textile Manufacturers Institute* v. *Donovan*, 452 U.S. 490 (1981)), rejected the use of cost-benefit analysis in setting OSHA health standards. However, the Court reaffirmed its previous position in the "benzene" case that a risk assessment is not only appropriate, but also required to identify significant health risk in workers and to determine if a proposed standard will achieve a reduction in that risk. Although the Court did not require OSHA to perform a quantitative risk assessment in every case, the Court implied, and OSHA as a matter of policy agrees, that assessments should be put into quantitative terms to the extent possible. The following paragraphs present an overall description of OSHA's preliminary quantitative risk assessment for occupational exposure to tuberculosis (TB).

An earlier version of this risk assessment was reviewed by a group of four experts in the fields of TB epidemiology and mathematical modeling. The reviewers were George Comstock, MD, MPH, DPH, Alumni Centennial Professor of Epidemiology, The Johns Hopkins University; Neil Graham MBBS, MD, MPH, Associate Professor of Epidemiology, The Johns Hopkins University; Bahjat Qaqish, MD, PhD, Assistant Professor of Biostatistics, University of North Carolina; and Patricia M. Simone, MD, Chief, Program Services Branch, Division of Tuberculosis Elimination, CDC. The reader is referred to the peer review report in the docket for additional details (Ex. 7-911). The revised version of OSHA's risk assessment, as published in this proposed rule, includes OSHA's response to the reviewers' comments as well as updated risk estimates based on recent purified protein derivative (PPD) skin testing data made available to the Agency since the peer review was

performed and is generally supported by the reviewers or is consistent with reviewers' comments. (Note: PPD skin test and tuberculin skin test (TST) are synonymous
terms.)

The CDC estimates that, once infected with *M. tuberculosis*, an untreated individual has a 10% lifetime probability of developing active TB and that approximately
half of those cases will develop within the first or second year after infection occurs.
Individuals with active TB represent a pool from which the disease may spread.
Based on data from the CDC, OSHA estimates that every index case (i.e., a person
with infectious TB) results in at least 2 other infections (Ex. 7-269). For some
percentage of active cases, a more severe clinical course can develop which can be
attributed to various factors such as the presence of MDR-TB, an allergic response
to treatment, or the synergistic effects of other health conditions an individual might
have. Further, OSHA estimates that for 7.78% of active TB cases, TB is expected
to be the cause of death. Section 6(b)(5) of the OSH Act states that,

> The Secretary, in promulgating standards dealing with toxic materials or harmful
> physical agents under this subsection, shall set the standard which most adequately
> assures, to the extent feasible, on the basis of the best available evidence, that no
> employee will suffer material impairment of health or functional capacity even if such
> employee has regular exposure to the hazard dealt with by such standard for the period
> of his working life.

For this rulemaking, OSHA defines TB infection as a "material impairment of
health", for several reasons. First, once infected with TB, an individual has a 10%
lifetime likelihood of developing active disease and approximately 1% likelihood
of developing more serious complications leading to death. Second, allergic reaction
and hepatic toxicity due to chemoprophylaxis with isoniazid, which is one of the
drugs used in the recommended course of preventive treatment, pose a serious threat
to a large number of workers. Third, defining infection with *M. tuberculosis* as
material impairment of health is consistent with OSHA's position in the Bloodborne
Pathogens standard and is supported by CDC and several stakeholders who participated in the pre-proposal meetings, as well as Dr. Neil Graham, one of the peer
reviewers of this risk assessment. In his comments to OSHA, Dr. Graham stated,

> The focus of OSHA on risk of TB infection rather than TB disease is appropriate.
> TB infection is a potentially adverse event, particularly if exposure is from a MDR-
> TB patient, or if the health-care or institutional worker is HIV seropositive. In addition,
> a skin test conversion will in most cases mandate use of chemoprophylaxis for >6
> months which is at least inconvenient and at worst may involve adverse drug reactions.
> (Ex. 7-271)

The approach taken in this risk assessment is similar to the approach OSHA
took in its risk assessment for the Bloodborne Pathogens standard. As with bloodborne pathogens, the health response (i.e., infection) associated with exposure to the
pathogenic agent does not depend on a cumulative level of exposure; instead, it is
a function of intensity and frequency of each exposure incident. However, unlike
hepatitis B, where the likelihood of infection once an exposure incident occurs is

known with some degree of certainty, the likelihood of becoming infected with TB after an exposure incident is not as well characterized. With TB, the likelihood of infection depends on the potency of an exposure incident and the susceptibility of the exposed individual (which is a function of the person's natural resistance to TB and his or her health status). Further, the potency of a given exposure incident is highly dependent on several factors, such as the concentration of droplet nuclei in the air, the duration of exposure, and the virulence of the pathogen (e.g., pulmonary and laryngeal TB are considered more infectious than other types).

The Agency has sufficient data to quantify the risk associated with occupational exposure to TB among health care workers in hospitals on a state-by-state basis. In addition to hospital employee data, OSHA has obtained data on selected health care employee groups from the TB Control Office of the Washington State Health Department. These groups include workers employed in long-term health care, home health care, and home care. Small entities are encouraged to comment and submit any data or studies on TB infection rates relevant to their business.

Because it is exposure to aerosolized *M. tuberculosis* that places workers at risk of infection, and not some factor unique to the health care profession, the Agency concluded that the experience of these groups of health care workers is representative of that of the other "high-risk" workers covered by this proposal. This means that the risk estimates calculated for these groups of workers are appropriate to use as the basis for describing the potential range of risks for workers in other work settings where workers can be expected to come into close and frequent contact with individuals with infectious TB (or with other sources of aerosolized *M. tuberculosis*) as an integral part of their job duties. As discussed in section IV (Health Effects), epidemiological studies, case reports, and outbreak investigations have shown that workers in various work settings, including but not limited to hospitals, have become infected with tuberculosis as a result of occupational exposure to aerosolized *M. tuberculosis* when appropriate infection control programs for tuberculosis were not in place.

In this preliminary risk assessment, OSHA presents risk estimates for TB infections, cases of active disease, and TB-related deaths (i.e., where TB is considered the cause or a major contributing cause of death) for workers with occupational exposure to tuberculosis.

A number of epidemiological studies demonstrate an increased risk of TB infection among health care workers in hospitals and other work settings. A brief review of a selection of these studies is presented below, followed by OSHA's estimates of excess risk due to occupational exposure. Finally, OSHA presents a qualitative assessment of the risk of TB infection caused by occupational exposure to tuberculosis in correctional facilities, homeless shelters, drug treatment centers, medical laboratories, and other high-risk work groups.

REVIEW OF THE EPIDEMIOLOGY OF TB INFECTION
IN EXPOSED WORKERS

There are several studies in the published scientific literature demonstrating the occupational transmission of infectious TB. Reports of TB outbreaks and epidemiologic surveillance studies have shown that health care and certain other workers are, as a result of their job duties, at significantly higher risk of becoming infected than the average person.

OSHA conducted a thorough search of the published literature and reviewed all studies addressing occupational exposure to tuberculosis and TB infection in hospitals and other work settings. All published studies show positive results (i.e., workers exposed to infectious individuals have a high likelihood of becoming infected with TB). Because there are so many studies, OSHA selected a representative subset of the more recent studies conducted in the U.S. to include in this section. These studies were chosen because they show occupational exposure in various work settings, under various working conditions, and under various scientific study designs.

OSHA's summary of the studies is presented in Table V–1(a) and Table V–1(b). These studies represent a wide range of occupational settings in hospitals, ranging from TB and HIV wards in high prevalence areas, such as New York City and Miami, to hospitals with no known TB patients located in low prevalence areas such as the state of Washington. The studies include prospective studies of entire hospitals or groups of hospitals, retrospective surveys of well-controlled clinical environments, such as an HIV ward in a hospital, and case studies of single-source infection (i.e., outbreak investigations).

Outbreak investigations describe occupational exposure to tuberculosis from single index patients or a well-defined group of patients. Such investigations are more likely to demonstrate an upper limit of occupational risk in different settings, usually under conditions of suboptimal environmental and infection controls. Although outbreak investigations demonstrate the existence of occupational risk under certain conditions and the importance of the early identification of suspect TB patients quite well, these studies do not provide information conducive to risk assessment estimations. Limitations of outbreak investigations include the frequent absence of baseline PPD test results, the difficulty of extrapolating the results to non-outbreak conditions of TB exposure, and, often, small sample sizes. Table V–1(a) lists some of the published outbreak investigations and shows the risks posed to health care workers by such outbreaks, as well as the failures in control programs contributing to these episodes.

Prospective and/or retrospective surveillance studies are used to estimate conversion rates from negative to positive in PPD skin testing programs. These conversion rates can be used to estimate the excess incidence of TB infection. Surveillance studies among health care workers lend themselves to a more systematic evaluation of the risk of TB infection than outbreak investigations, for several reasons. First, these studies better reflect the risk of TB experienced by workers under routine conditions of exposure. Second, these studies are usually based on a larger group of workers and therefore yield more precise and accurate estimates of the actual risk of infection. However, the extent to which results from surveillance studies can be generalized depends on a careful evaluation of the study population. Some studies report skin test conversion rates for all workers in the hospital(s) under study. Such studies often include large groups of employees with little or no exposure to TB. Results from such studies may reflect an overall estimate of risk

TABLE V–1(A)
Outbreak Investigations of TB Infection

Authors/year	Setting/source	Risk of TB in health care workers	Contributing factors
Catanzaro (1982)	Hospital intensive care unit/San Diego/1 index case—7-day hospital stay.	14/45 (31%) PPD conversions, 10/13 (77%) PPD conversions among health care workers present at bronchoscopy.	Poor ventilation. No report on respirator use.
Kantor et al. (1988)	VA hospital in Chicago autopsy room/1 index case undiagnosed until histology exam of autopsy tissue.	9/56 (16%) PPD conversions among exposed workers vs. 3/333 (1%) conversions among unexposed (RR=17.8) 3 workers developed active TB.	No mechanical ventilation on medical ward (autopsy room): no isolation. Autopsy room had 11 air changes/hour and no air recirculation.
Beck-Sague (1992)	Jackson Memorial Hospital in Miami MDR-TB in HIV/patients on HIV ward and clinic during 1989-91.	13/39 (33%) PPD conversions on HIV ward and clinic.	Some rooms had positive pressure. Inadequate triage of patients with suspected TB. Delay in use of isolation. Early discharge from isolation.

in that environment, but may underestimate the occupational risk of those with frequent exposure.

Other surveillance studies report PPD conversion rates of more narrowly-defined groups of workers, usually those working in "high-risk" areas within a hospital such as the HIV or TB wards. Some of these studies have internal control groups (i.e., they compare PPD conversion rates between a group of workers with extensive exposure to TB and a group of workers with minimal or no exposure to TB), thus making it possible to more precisely quantify the magnitude of excess risk due to occupational exposure. However, these studies are also limited in their usefulness for risk assessment purposes. They usually have small sample sizes, making it more difficult to observe statistically significant differences. More importantly, internal control groups may overestimate background risk, and thus underestimate excess occupational risk, unless painstaking efforts are made to eliminate from the control group those individuals with the potential for occupational exposure, a difficult task in some hospital environments. Table V–1(b) contains a selected list of published surveillance studies.

In reviewing Table V–1(a) and Table V–1(b), the reader should bear in mind that these tables are not intended to present an exhaustive list of epidemiologic studies with TB conversion rates in occupational settings. Instead, these tables present brief summaries of some of the epidemiologic evidence of occupational TB transmission found in the published literature; they are intended to convey the seriousness of the risk posed to health care workers and to illustrate how failures in

TABLE V–1(B)
Surveillance Studies of TB Infection in Exposed Health Care Workers

Authors/year	Setting/source	Study period	Population	Risk of TB in health care workers	Comments
Price et al. (1987)	19 Eastern North Carolina hospitals. 29 Central North Carolina hospitals. 8 Western North Carolina hospitals.	1980-84	All Hospital workers	1.80% annual PPD conversion rate. 0.70% annual PPD conversion rate. 0.61% annual PPD conversion rate.	
Aitken et al. (1987)	64 hospitals in Washington State.	1982-84	All Hospital workers	0.1% PPD conversion rate/in 3 years.	Strict adherence to CDC guidelines.
Malasky et al. (1990)	14 urban hospitals in U.S.	(¹)	Physicians in training in pulmonary medicine and infectious disease.	11% PPD conversion/3 years among pulmonary fellows, 2.4% PPD conversions/3 years among infectious disease fellows.	
Dooley et al. (1992)	Hospital in Puerto Rico TB in HIV-infected patients.	1989-90	Hospital workers (n=908)	Prevalence study: 54/109 (50%) nurses exposed to TB patients had positive PPDs 35/188 (19%) clerical workers with no exposure to TB had positive PPDs (p<0.001).	Isolation rooms did not have negative pressure. Recirculated air was not filtered.
NIOSH	Jackson Memorial Hospital, Miami.	1989-92	Hospital workers in selected wards (n=607).	60% annual PPD conversion among 263 exposed workers, 0.6% annual PPD conversion among 344 unexposed workers.	Incomplete isolation facilities. Improper application of isolation procedures.
Cocchiarella et al. (1996)	Cook County Hospital, Chicago.	1991	Graduating physicians with at least 1 year of clinical work at CCH (n=128).	18.8% 3-year PPD conversion rate for house staff in internal medicine vs. 2.2% PPD conversion rate for house staff in other specialties.	Residents were offered limited respiratory protection during exposures. No protocol available for early identification of suspect TB cases. PPD testing program incomplete. Inadequate isolation facilities.

¹ Mid 1980's (3 years).

control programs contribute to this risk. Upon reviewing these studies, a consistent pattern emerges: these work settings are associated with a high likelihood for occupational exposure to tuberculosis, and high rates of TB infection are being observed among health care workers.

QUANTITATIVE ASSESSMENT OF RISK

Data availability usually dictates the direction and analytical approach OSHA's risk assessment can take. For this rulemaking, three health endpoints will be used: (1) TB infection, which is "material impairment of health" for this proposed standard; (2) Active disease following infection; and, (3) Risk of death from active TB.

In order to account for regional variability in TB prevalence and therefore to account for expected variability in the risk of TB infection in different areas, the Agency chose to develop occupational risk estimates on a state-by-state basis. This approach was criticized by Dr. Neil Graham as being too broad and "* * * insufficient in light of the tremendous variability * * * that can occur within a state." (Ex. 7-911). The Agency recognizes that risk estimates on a county-by-county basis would be preferable; however, the unavailability of comprehensive county data has prevented the Agency from conducting such analysis.

The annual excess risk of TB infection due to occupational exposure is defined as a multiplicative function of the background rate of infection and is expressed as:

$$p = ERR_o * R_b$$

where:
p is the annual excess risk due to occupational exposure,
R_b is the background rate of TB infection, and
ERR_o is a multiplicative factor denoting the excess relative risk due to occupational exposure (ERR_o).

Estimates of ERR_o are derived from surveillance studies of workers with occupational exposure to TB. ERR_o is defined as the relative difference between the overall exposed worker risk and the background (population) risk and is calculated as the difference between overall worker and background risk divided by the background risk.

The annual excess risk due to occupational exposure is defined as a function of the background risk because of data limitations. If data on overall worker risk were available for each state, then the excess risk due to occupational exposure would simply be the difference between overall worker risk and background risk. Instead, the annual excess risk due to occupational exposure (i.e., p) is estimated using a multiplicative model because data on overall worker risk (i.e., R_w) were available only for the states of Washington, and North Carolina and for Jackson Memorial Hospital located in Miami, Florida. Therefore, the annual excess risk due to occupational exposure in state i (p_i) is expressed as:

$$p_i = \frac{\left(R_{wj} - R_{bj}\right)}{R_{bj}} * R_{bi}$$

where:
R_w is the overall worker risk estimated from surveillance studies (study j),

R_{bj} is the study control group risk (i.e., study background risk), and
R_{bi} is the background rate for state i.

When i=j (i.e., Washington State or North Carolina), the excess risk due to occupational exposure, is expressed as the straight difference between overall worker risk and background risk.

OSHA calculated estimates of ERR_o based on three occupational studies: the Washington State study, the North Carolina study, and the Jackson Memorial Hospital study (Exs. 7-263, 7-7, 7-108). These estimates were expressed as percent change above each study's background. The derivation of these estimates is described in section 2.

In order to estimate an overall range of occupational risk of TB infection, taking into account regional differences in TB prevalence in the U.S., OSHA: (1) Estimated background TB infection rates by state (R_{bi}), and (2) applied estimates of ERR_o, derived from the occupational studies, to the state background rates to calculate estimates of excess risk due to occupational exposure by state.

OSHA used a multiplicative function of each state's background infection rate to estimate excess risk of TB infection because the probability of occupational infection can be viewed as a function of the number of contacts and frequency of contacts with infectious individuals. Thus, estimates of expected relative increase in risk above background due to occupational exposure are calculated for the three available studies and these relative increases (i.e. ERR_o) are multiplied by background rates for each state to derive estimates of excess occupational risk by state. These state estimates are then used to derive a national estimate of occupational risk.

The CDC compiles and publishes national statistics on the incidence of active TB in the U.S. by state based on reported cases. OSHA relied on these data to estimate TB infection background rates through the use of a mathematical model because information on TB infection is not being collected nationwide by CDC. A more detailed discussion on the methodology and derivation of background risk estimates by state is found in section 3, and discussion on the estimation of occupational risk estimates by state is found in section 4 of this risk assessment.

Because section 6(b)(5) of the OSH Act requires OSHA to assess lifetime risks, OSHA has converted the annual excess risk due to occupational exposure into an excess lifetime risk based on a 45-year working lifetime. The formula used to calculate lifetime occupational risk estimates of the probability of at least one occurrence of TB infection due to occupational exposure in 45 years is expressed as $\{1-(1-p)^{45}\}$, where p is the annual excess risk due to occupational exposure. Two assumptions are critical in defining lifetime risk: (1) the exposure period is 45 years, and (2) the annual excess risk remains constant. The implication of the second assumption is that the worker's exposure profile and working conditions, which may affect the level and intensity of exposure, and the virulence of the pathogen, remain unchanged throughout a working lifetime. The merit of this assumption was questioned by Dr. Graham, because, as he states "* * * patient contact may vary greatly throughout a career for many HCWs [health care workers]." and "* * * physicians (and nurses) often do not have extensive patient contact until [their] mid-twenties, while other workers increasingly retire early." Dr. Graham recommends that OSHA's risk assessment be adjusted to account for variable exposure levels and variable working lifetimes. Although accounting for variable exposure levels could result in more precise risk estimates, the unavailability of comprehensive information on lifetime TB exposure scenarios by occupational group prevented the Agency from developing a more complex risk model.

OSHA has customarily assumed a 45 year working lifetime in setting health standards. The Agency believes that this assumption is reasonable and consistent with the Act. The Act requires the Secretary to set a standard for toxic substances that would assure "no employee * * * suffer material impairment of health or functional capacity *even if such employee has regular exposures to the hazard for the period of his working lifetime."* 29 U.S.C. § 655(b)(5) (emphasis added). The U.S. Court of Appeals for the District of Columbia upheld the use of a 45-year lifetime in the asbestos standard against an assertion by the Asbestos Information Association that the average duration of employment was five years. *Building and Construction Trades Department, AFL-CIO* v. *Brock*, 838 F.2d 1258, 1264, 1265 (D.C. Cir. 1988). The Court said that OSHA's assumption "appears to conform to the intent of Congress" as the standard must protect even the rare employee who would have 45 years of exposure. Id. at 1264. In addition, while working lifetimes will vary, risk is significant for some who work as little as one year and, at any rate, individual and population risks are likely to remain the same so long as employees who leave one job are replaced by others, and those who change jobs remain within a covered sector. Nevertheless, the Agency solicits information regarding the likelihood of exposure to active TB in the workplace and duration of employment in various occupational groups. Lifetime risk estimates of TB infection by state are described in section 4.

Lifetime risk estimates of developing active TB are calculated from lifetime risk estimates of TB infection assuming that, once infected, there is a 10% likelihood of progressing to active TB. These estimates are discussed in section 4. Further, the number of deaths caused by TB is calculated from the lifetime estimates of active TB using OSHA's estimate of TB case fatality rate, also discussed in section 4.

Definitions

For the purpose of estimating incidence rates, *TB infection* rate is defined as the annual probability of an individual converting from negative to positive in the tuberculin skin test. *Annual occupational* risk is defined as the annual excess risk of becoming infected with TB due to occupational exposure, and is estimated as a function of the background risk. *Lifetime occupational risk* is defined as the excess probability of becoming infected with TB due to exposure in the workplace, at least once, in the course of a 45-year working lifetime and is estimated as $\{1-(1-p)^{45}\}$ where p is the annual occupational risk of TB infection.

Data Sources for Estimating Occupational Risk

The quantitative data needed to develop an overall national estimate of risk for TB infection due to occupational exposure are not available. The CDC does not publish occupational data associated with TB infection incidence and active TB on a nation-wide basis. There has been some effort to include occupational information on the TB reporting forms, but only a limited number of states are currently using the new forms that capture occupational information in a systematic way.

However, there are a number of sources that permit the risk in occupational settings to be reasonably estimated and, with the aid of mathematical models, to develop estimates of excess relative occupational risk (ERR_o), which can then be multiplied by the state-specific background rates to yield estimates of excess occupational risk. OSHA has identified three data sources that are suitable for assessing the excess risk of TB infection in health care workers with occupational exposure. These include: (1) A 1994 survey of tuberculin skin testing in all health care facilities in Washington State; (2) A state-wide survey of hospitals in North Carolina, conducted in 1984-1985, which addressed TB skin testing practices, TB infection prevalence, and TB infection incidence among hospital employees in that state; and (3) the employee tuberculin skin test conversion database from Jackson Memorial Hospital in Miami, Florida. In addition to these hospital employee data, the Agency has obtained data on selected other work groups from the state of Washington. These groups include workers employed in long-term health care, home health care, and home care.

On the issue of data availability for this risk assessment, Dr. Graham agrees with OSHA that there are no comprehensive data available with respect to occupational risk of TB infection in health care and other institutions in the U.S. Instead of relying on two state specific studies, Dr. Graham recommends, though with serious reservations, the use of a review study by Menzies *et al.* (Ex. 7-130). Dr. Graham admits that the "validity of the estimates in these reports [reviewed in the Menzies *et al.* study] must be open to serious question * * *" for the following reasons, which were pointed out by Dr. Graham: several of the studies reviewed are very old and not relevant to TB risk in the 1990s; four studies use tine tests and self-reports of skin test results, which are not useful for estimation of risk of TB infection; the studies were not consistent in the inclusion of high and low risk workers; two-step testing was not done; and the participation rates were extremely low or unreported in many of the studies included in this review.

OSHA has chosen not to rely on the Menzies *et al.* review study, because, in addition to Dr. Graham's reservations (which the Agency shares), OSHA is also concerned about the inclusion in the Menzies *et al.* review article of studies conducted outside the U.S. Factors known to affect the epidemiology of TB, such as environmental conditions, socio-economic status, and work practices, are expected to differ greatly from one country to another, and are not controlled for in the statistical analyses of these studies. For all of these reasons, the Agency has chosen to rely solely on U.S. studies for its quantitative risk estimations.

Estimates of excess risk due to occupational exposure are expressed as the percent increase above background based on relative risk estimates derived from occupational studies. Internal control groups provided estimates of background risk for the Washington state and Jackson Memorial data sets. In the absence of a suitable internal control group, the estimated annual state-wide TB infection rate, as calculated in Section 3, was used as the background rate in the North Carolina study.

TABLE V–2
Washington State 1994 Survey Results

Type of facility	Number of [a] establishments	Number of skin tests	Number of conversions	Annual rate of TB conversion
Hospital	76 (85%)	39,290	50	1.27/1,000
Long-term Care	142 (81%)	11,332	111	9.80/1,000
Home Health Care	47 (77%)	2,172	11	5.06/1,000
Home Care	8 (80%)	537	1	1.86/1,000
Total	273 (81.5%)	53,331	173	3.24/1,000

[a] Numbers in parentheses are study response rates for each group.

Washington State Data

Initially, OSHA relied on a three-year prospective study, conducted between 1982 and 1984 in the state of Washington, to derive an estimate of excess risk for TB infection as a result of occupational exposure (Ex. 7-42). OSHA received several objections to the use of this study. The study used hospitals with no known TB cases as "controls" based on the assumption that in those hospitals the risk of TB infection to employees may be the same as for the general population. Dr. Qaqish noted that this assumption is highly questionable and that the use of such controls is not appropriate. Dr. Graham and Dr. Qaqish pointed out that the published results did not include conversions identified through contact investigations, which means that the conversion rate reported in that study was likely to be an underestimate of the true risk. In addition, the commenters noted that the study was designed to estimate the effectiveness of the TB screening program and may have produced skin testing results biased toward the null; the study is relatively old; and, the study was conducted prior to the AIDS epidemic and therefore the results may not be relevant to the occupational risk at present because the relationship between HIV and TB is not reflected in this study.

In an effort to respond to reviewers' comments, the Agency chose to update the analysis by relying on a data set of tuberculin skin testing results from a survey of the state's tuberculin skin testing program in 1994. This survey is conducted by the TB Control Office in the Washington State Health Department and it covers all hospitals in the state, as well as long-term care, home health care, and home care facilities. OSHA was given access to the database for the 1994 survey as well as data on conversions identified through contact investigations for the same year (Ex. 7-263). Table V–2 summarizes the results of the 1994 survey. Of the 335 health care establishments in the state of Washington, 273 responded to the survey, for an overall response rate of 81.5%. Of those, 76 were hospitals, 142 were long-term care, 47 were home health care, and 8 were home care facilities. Hospitals had the highest survey response rate (85%) and home health care had the lowest (77%). Every employee at risk for TB infection (i.e., who was known to be tuberculin skin test negative at the start of the study period) in the participating hospitals and long-term care facilities was given a tuberculin skin test, including administrators, housekeepers, business office staff, and all part-time employees. Testing in home health care facilities was generally confined to those nursing staff who had direct client contact. Employees in home care are those who provide services to patients in home health care and include food handlers, cleaning aides, personal care-givers, and some social workers.

The overall rate of skin test conversion for workers in the health care system in Washington State in 1994 was 3.24 per 1,000 employees tested. This is greater than a 4-fold increase from the estimated state-wide background rate of 0.69 per 1,000 at risk, as calculated in section 3. The annual rate of TB conversion ranged from 1.27 per 1,000 tested for hospital employees to 9.80 per 1,000 tested for long-term care employees.

The annual rate of 9.8 per 1,000 for long-term care employees probably reflects the high potential for exposure to undiagnosed active TB in those facilities. As a rule, long-term facilities in Washington State do not have AFB isolation rooms. Therefore, residents with no obvious TB symptoms but who might be infectious spend most of their time in open spaces exposing other residents and workers to infectious droplet nuclei. However, once a resident has been identified as a suspect TB patient, that person is transferred to a hospital until medically determined to be non-infectious.

Also, since employees who were 35 years of age or younger were not given a two-step test at hiring, and a high percentage of employees are foreign born and therefore most likely to have been vaccinated during childhood with the BCG vaccine, some of the conversions observed might be late boosting because of BCG. However, an almost two-fold increase in risk for long-term care workers even as compared to the significant excess risk among home health care workers clearly indicates that the risk of TB infection for workers in long-term care is high and not likely to be fully explained by late boosting. Beginning in 1995, two-step testing has been done on all new hires in Washington State. Thus, tuberculin skin testing data for 1995 are not expected to be influenced by possible late boosting; OSHA will place the 1995 data in the rulemaking record as they become available.

Hospital workers had the lowest overall rate of conversion (overall rate of 1.27 per 1,000). This, in part, can be attributed to the existence of extensive TB control measures in that environment in Washington State. Compliance with the CDC Guidelines and OSHA's TB Compliance Directive is quite high in Washington State because: (a) There is a strong emphasis on early identification of suspect TB patients; (b) there is a strong emphasis on employee training and regular tuberculin skin testing (although on a less-frequent basis than recommended in the Guidelines: All employees are tested at hire and annually thereafter); (c) the use of respirators is expected when entering an isolation room; and (d) all isolation rooms are under negative pressure, have UV lights, and exhaust to the outside. In addition, conversion data in hospitals are more likely to represent true TB infections than in the other health care settings, because hospitals are more likely to re-test converters in an effort to eliminate false-positive cases.

A more thorough analysis of the hospital data is presented in table V–3. Because the Washington State survey was not designed to compare exposed persons with matched controls who have had no exposure, several alternative definitions of an internal control (unexposed) group were used in analyzing this data set. Three

TABLE V–3
Washington State Data Hospital PPD Skin Testing Results

Definition of exposed and control groups	Sample size	Number of skin tests given	Number of conversions observed	Average conversion rate 1[a]	Overall conversion rate 2[b]	Relative Risk Rate 1	Relative Risk Rate 2
Definition 1							
Control: Hospitals in zero-TB counties and with no-known TB patients	16	1,142	1	0.477	0.8756		
Exposed: Hospitals in counties reporting TB or having TB patients	60	38,148	49	1.523	1.28447	3.19	1.47
Definition 2							
Control: Hospitals that transfer out TB patients	35	3,645	3	0.498	0.823		
Exposed: Hospitals with isolation rooms	41	35,645	47	1.989	1.3185	3.99	1.602
Definition 3							
Control: State-wide estimates of annual risk of infection				[c]0.69	[c]0.69		
Exposed: All PPD testing data	76	39,290	50	1.302	1.27	1.89	1.84

[a] Rate 1 is estimated as the arithmetic average of hospital specific conversion rates.

[b] Rate 2 is estimated as the ratio of the sum of all conversions reported divided by the total number of skin tests given within each group.

[c] Source: Table V–3(b), state-wide rate of infection.

different analyses, shown in table V–3, produced estimates of annual occupational infection rates ranging from 0.4 to 0.6 per 1,000 above control (i.e., ranging from a 47% to an 84% increase above control). In order to minimize the likelihood of contaminating the control group with persons having significant occupational exposure, OSHA defined the control group as workers in hospitals located in zero-TB counties and with no known TB patients. This analysis is summarized in table V–3 as Definition 1. If potential for occupational exposure is defined as either working in a hospital in a county that has active TB or in a hospital that has had TB patients, then the annual risk due to occupational exposure is 47% above background. The excess annual risk due to occupational exposure appears to be approximately 60% above background, if workers in hospitals with a transfer-out policy for TB patients are considered to be the control group, shown as Definition 2 in table V–3. A 60% increase above background is not statistically significantly different from a 47% increase and therefore these two "control" groups can be viewed as producing "statistically" equivalent results. However, the Agency believes that Definition 1 is more appropriate, though the risk estimates are higher if the control group is defined based on Definition 2, because there is a higher likelihood of potential for exposure to a patient with undiagnosed TB under Definition 2 conditions. Comparisons of all hospital TST data to the state-wide estimate of TB infection rate resulted in an estimate of the annual excess occupational risk of approximately 84% above background, shown in table V–3 as Definition 3. Estimates of the annual and lifetime occupational risk of TB infection for the average health care worker in hospitals by state, extrapolated from this study and using Definition 1 as the control group, are presented and summarized in section 4.

Annual rates of excess risk due to occupational exposure were estimated for long-term care, home health care, and home care and are presented in Section 4. The same control group used in the hospital data analysis, Definition 1 (i.e., 0.876/1,000 workers at risk) was used to estimate the background risk among workers in long-term care, health care, and home care facilities and settings. Using 0.876 as the background infection rate for workers in these settings (a) provided a level of consistency among the Washington data analyses, and (b) resulted in a lower estimate of occupational risk for the non-hospital health care workplaces than would have resulted had the state-wide background risk estimate (i.e., 0.67/1,000 see Section 3) been used. When industry-specific risk data are used, there is approximately a 10-fold increase in annual risk for workers in long-term care, a 5-fold increase in annual risk for workers in home health care, and a 1-fold increase in annual risk for workers in home care (see Section 4).

Estimates of the range of annual and lifetime occupational risk for the average health care worker in long-term care, home health care, and home care by state, extrapolated from the Washington State study, are presented in Section 4.

North Carolina Study

A state-wide survey of all hospitals in North Carolina (NC) was conducted in 1984-1985 (Ex. 7-7). The survey's questionnaire was designed to address three main areas of concern affecting hospital employees: (1) Tuberculin skin testing practices; (2) TB infection prevalence; and (3) TB infection incidence. The incidence of new infections among hospital personnel was assessed over a five-year period by reviewing tuberculin skin test conversion data during calendar years 1980 through 1984 and was calculated as the number of TB skin test conversions divided by the number of skin tests administered. (Since most employees were only given annual testing, the number of tests administered is a very close estimate of the total number of people tested within a year and thus can be used as the denominator in estimating infection incidence.) Only 56 out of 167 hospitals reported information on TB conversion rates (34% response rate). The authors estimated a state-wide TB infection rate of 11.9 per 1,000 per year for hospital employees in 1984 and a five-year mean annual infection rate of 11.4 per 1,000, with a range of 0-89 per 1000 employees at risk for TB infection. An analysis of the data by region (i.e., eastern, central, western) showed that the eastern region had consistently higher rates (with an average infection rate of 18.0 per 1,000) followed by the central region (7.0 per 1,000) and the western region (6.1 per 1000). Results of this study are shown in table V–4.

Use of this study's overall results for risk estimates was criticized by the peer reviewers because of design flaws in the study (e.g., high non-response rate, inconsistent

TABLE V–4
Skin Test Conversion Rates[a] North Carolina Hospital Personnel[b]

Region	Year					
	1980	1981	1992	1993	1984	5-year mean
Eastern	19.3	30.8	17.7	11.2	15.7	18.0
	(7)	(10)	(11)	(12)	(18)	(19)
Central	3.0	3.7	7.2	6.6	10.0	7.0
	(6)	(8)	(13)	(23)	(25)	(29)
Western	1.9	13.5	5.3	4.1	7.2	6.1
	(2)	(4)	(4)	(4)	(8)	(8)

[a] Conversion rates are expressed as number of conversions per 1,000 workers tested.
[b] In parentheses is the number of hospitals included in the study.

skin testing practices, and limited two-step testing) and, most importantly, the presence of atypical mycobacteria (contributing to false positive results) in the eastern part of the state. Based on further input from Dr. Comstock, the Agency chose to rely on the study results from the western region only, because they are considered to be more representative of the "true" risk of infection and are expected to be less confounded by cross-reactions to atypical mycobacteria. Further, the Agency chose to rely on the conversion rate estimated for 1984 because it was the most recent data reported in the study. Therefore, the western region conversion rate of 7.2 per 1,000, estimated based on responses to the survey from eight hospitals in 1984, was used as an overall worker conversion rate. Further, the 1984 rate was adjusted by the percent decrease of active TB between 1984 and 1994 in North Carolina so that the final worker conversion rate for 1994 based on the western region rates reported in this study was estimated to be 5.98 (7.2 * 532/641 = 5.98) per 1,000 employees at risk for TB infection.

The North Carolina study did not have an internal control group to use as the basis for estimating excess risk due to occupational exposure because the conversion rates presented in this study were based on TST results for the entire hospital employee population. In the absence of an internal control group, the Agency used the estimated state-wide background rate of 1.20 per 1,000 as the background rate of infection for the western region in North Carolina (see Section 3) to estimate excess risk due to occupational exposure.[1] Based on this study, annual occupational risk is approximately four times greater than background [(5.98–1.2)/1.2 = 3.98]. Estimates of the annual and lifetime occupational risk of TB infection based on this study by state are presented in Section 4.

---

[1] Using the state-wide estimate of population risk as the background estimate of risk for this study most likely results in an underestimate of the true excess risk due to occupational exposure, because the true background estimate of risk for the western region in North Carolina is expected to be less than the state-wide estimate, which is influenced by the large number of infections found in the eastern region of that state.

Jackson Memorial Hospital Study

Jackson Memorial Hospital (JMH) is a 1500-bed general facility located in Miami, Florida, employing more than 8,000 employees. It is considered one of the busiest hospitals in the U.S. It is the primary public hospital for Dade County and the main teaching hospital for the University of Miami School of Medicine. JMH treats most of the TB and HIV cases in Dade County and, consequently, there is a higher likelihood of occupational exposure to TB in this facility than in the average hospital in the U.S. From March 1988 to September 1990, an outbreak of multidrug-resistant TB (MDR-TB) occurred among patients and an increased number of TST conversions was observed among health care workers on the HIV ward. This prompted a re-evaluation of the hospital's infection control practices and the installation of engineering controls to minimize exposure to TB. As part of the evaluation of the outbreak, NIOSH did a Health Hazard Evaluation and issued a report (Ex. 7-108). In addition, NIOSH conducted a retrospective cohort study of JMH to determine whether the risk of TB infection was significantly greater for health care workers who work on wards having patients with infectious TB than those who work on wards without TB patients.

For the data analysis of this study, "potential for occupational exposure" was defined based on whether an employee worked on a ward that had records of 15 or more positive cultures for pulmonary or laryngeal TB during 1988-1989. In other words, positive culture was taken as a surrogate for exposure to infectious TB. The authors restricted the "exposed" group to employees on wards with exposures to pulmonary or laryngeal TB because they intended to restrict the study to hospital workers with exposure to patients with the highest potential for being infectious. There were 37 wards at JMH that had submitted at least one positive culture during 1988-1989. Seven wards met the criteria of 15 or more and were therefore included in the "exposed" group. These were the medical intensive care unit, five medical wards, and the emergency room. The "control" group was defined as hospital workers assigned to wards with no TB patients (i.e., wards with no records of positive cultures during 1988-89). The "control" wards were post-partum, labor and delivery, newborn intensive care unit, newborn intermediate care unit, and well newborn unit. The results of this analysis are presented in Table V–5.

Table V–5 shows a substantially elevated risk for those workers with potential exposure to patients with infectious TB. The relative risk ranges from 9 to 11.7 between 1989 and 1991 and is statistically significant for all of those years. This suggests that the excess risk due to occupational exposure is approximately 8-fold above background; this is an overall risk estimate that reflects the occupational risk of TB infection for JMH employees with patient contact, because this analysis included everyone tested in the "exposed" and "control" group, regardless of his or her specific job duties or length of patient contact.

An analysis of various occupational groups within this cohort showed that nurses and ward clerks in the "exposed" groups had the highest conversion rates: 182 and 156 conversions per 1,000 workers tested, respectively. Other studies have shown that health care workers who provide direct patient care are at greater risk for infection than workers who do not provide direct patient care. The high risk seen in ward clerks was unexpected since these workers are not involved in direct patient

TABLE V–5
Skin Test Conversion Rates for Hospital Personnel at
Jackson Memorial Hospital[a,b]

Year	Exposed group	Control group	Relative risk	95% confidence interval
1989	62.2 (13/209)	6.2 (2/324)	10.1	2.3–44.2
1990	75.5 (16/212)	6.5 (2/309)	11.7	2.7–50.2
1991	31.7 (6/189)	3.5 (1/282)	9.0	1.1–73.8

[a] Rates are expressed as number of conversions per 1,000 workers tested.
[b] Source: Ex. 7-108

care. However, in the emergency room, the risk for TST conversion for the ward clerks was almost three times higher than for the nurses, 222 and 83 per 1,000, respectively. Ward clerks in the emergency room are responsible for clerical processing of patients after triage, handling specimens for the laboratory, and gathering clothing and valuables from admitted patients. During these interactions, there may have been less strict adherence to infection control measures, and this could explain the high conversion rate.

OSHA used the results from the 1991 analysis of the data in the JMH study to estimate occupational risk of TB infection in hospital workers with a relatively high likelihood of occupational exposure, for the following reasons: (a) 1991 represents the most recent year for which conversion data are available prior to the time when TB infection control measures were fully implemented at JMH; and (b) The higher conversion rates reported for 1990 and 1989 (75.5 and 62.2 per 1,000 respectively) may be atypical, i.e., they may to some extent reflect the effect of the outbreak and not the long-term occupational risk.

Based on the results of this study, OSHA estimates that the annual excess risk of TB infection due to occupational exposure is 7.95 times greater than background. Estimates of annual and lifetime occupational risk of TB infection for the average health care worker in hospitals by state, extrapolated from this study, are presented and summarized in section 4.

Estimation of Background Risk of TB Infection

OSHA's methodology for estimating population (background) TB infection rates relies on the assumption that TB infection occurring in an area can be expressed as a numerical function of active TB cases reported in the same area. If the likelihood of observing any infection in a population is minimal, then the likelihood of observing active disease diminishes. Conversely, the presence of active TB implies the presence of infection, since active disease can only progress from infection. Therefore, there is a functional relationship linking TB infections to active disease being observed in a particular area during a specified time period.

Peer reviewer comments on this assumption varied. Neil Graham states in his comment "Although factors such as migration and distribution of the population may influence this relationship it seems probable that this assumption is largely correct and justifiable." (Ex. 7-271). On the other hand, Dr. Simone expresses concern over this assumption and states "It is not necessarily true that a change in cases now reflects the risk of infection now." Dr. Qaqish demonstrates in his comment that the net effect of assuming a proportional relationship between the number of active cases and the number of new infections is to introduce a possible bias into the estimate of background risk of TB infection, although such a bias could work in either direction, i.e., toward increasing or decreasing the estimate of risk. Dr. Qaqish further states that in the absence of more "relevant data," it is not possible to determine the actual net effect in magnitude and direction of the bias and "without obtaining additional data, it would be impossible for the Agency to improve on the accuracy of the risk estimates * * *" OSHA has considered all of the reviewer comments and is aware of the inherent uncertainty and the potential for bias associated with the use of this assumption; however, in the absence of the additional "relevant" data to which Dr. Qaqish refers, the Agency believes this approach to be justifiable.

In defining the model used to estimate the annual infection rates occurring in a geographical area based on data on active disease cases reported for the same area, infections progressing to active disease are assigned to one of three distinct groups: those occurring this year, last year, and in previous years.

TB cases reported to CDC each year are a combination of new and old infections that have, for various reasons, progressed to active disease. Until recently, it was believed that most of the active cases were the product of old infections. However, with the use of DNA fingerprinting techniques, researchers have reported that a larger percentage of active cases may be attributed to new or recent infections. Small *et al.* reported, in an article on tracing TB through DNA fingerprinting, that as many as 30% of the active cases reviewed in the study may be the result of recent infections (Ex. 7-196).

In this risk assessment, the Agency assumes the lifetime risk that an infection will progress to active TB to be approximately 10%. This estimate is supported by CDC and in her comment, Dr. Simone states that: "The assumption * * * is generally agreed upon." Dr. Comstock and Dr. Qaqish both questioned the validity and accuracy of CDC's estimate. Their comments suggest that the true lifetime rate of progression from infection to active disease for adults may be less than 10 percent. However, as Dr. Graham points out, the 10% assumption is a widely accepted "rule of thumb" and is also in relative agreement with data from the unvaccinated control group of the British Medical Research Council (MRC) vaccination trial in adolescents (Ex. 7-266).

In the MRC study, 1,338 adolescents' skin tests converted following TB exposure where the precise date of conversion was known. Of these, 108 (8.1%) individuals developed active TB during follow-up. Of these, 54% developed active TB within one year and 78% within 2 years. This results in a risk of approximately 4% at one year, 6% at two years, and an overall risk of 8%. Given that the risk of TB reactivation increases with age, the lifetime risk is expected to be higher than the 8% attained in this study and, as Dr. Graham points out, a 10% overall lifetime risk seems reasonable.

Based on Dr. Graham's recommendation to rely on the progression rates from the MRC study, OSHA changed the assumption on the progression parameters from 2.5% (first year), 2.5% (second year), and 5% (remaining lifetime) to 4%, 2% and 4%, respectively. Therefore the total 10% progression from infection to active disease is partitioned into 3 groups: progression during the first year after infection (40% of all infections that eventually progress, for a net probability of 4%), progression during the second year (20% of all infections that eventually progress, for a net probability of 2%), and progression during all subsequent years (the remaining 40% of progressing infections). This last probability (4%) is assumed to be uniformly distributed across the remaining lifespan.

TB rates vary considerably by geographic area, socio-economic status, and other factors. In an attempt to account for some of those factors, to the extent possible, background TB infection rates have been estimated separately for each state. The derivation of background infection rates involves several steps for which the process and formulae are presented below.

Step 1: Background rate of TB infection for state i in year j is defined as:

$$B_{i(j)} = I_{i(j)}/X_{i(j)} \tag{1}$$

where:

$B_{i(j)}$ is the background TB infection rate for state i in year j
$I_{i(j)}$ is an estimate of the number of new infections that occurred in state i in year j
$X_{i(j)}$ is the population at risk for TB infection in state i in year j.

Step 2: Estimation of $I_{i(j)}$, the number of new TB infections:
Let:
$A_{i(j)}$ be the total number of adult TB cases reported to CDC by state i in year j.
$A_{(j)}$ be the total number of adult TB cases reported to CDC by all states in year j.
$P_{i(j)}$ be the estimated prevalence of adult TB infection in state i during year j.
R_i be the ratio of the number of adult TB cases reported in 1993 to the number of adult cases reported in 1994 in state i.

The number of TB cases reported in 1994 can be expressed as a function of TB infections expected to have progressed to active disease, by the following formula:

$$A_{i(1994)} = .04 * I_{i(1994)} + .02 * I_{i(1993)} + (.04/73) * I_{i(1992)} * prob(\text{alive in 1994})$$

$$+ (.04/73) * I_{i(1991)} * prob(\text{alive in 1994})$$

$$+ ....$$

$$+ ....$$

$$+ (.04/73) * I_{i(1919)} * prob(\text{alive in 1994})$$

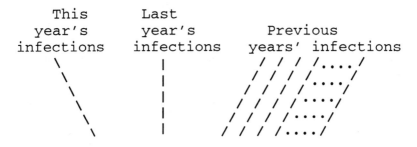

This year's active TB cases

This can be expressed as:

$$A_{i(1994)} = .04 * I_{i(1994)} + .02 * I_{i(1993)} + (.04/73) * \Sigma [I_{i(j)} * \text{prob(alive in 1994)}],$$

where j ranges from 1919 to 1992. The quantity inside the summation symbol is the sum of all people who were infected with TB between 1919 and 1992 and are still alive in 1994. This summation can be approximated by the prevalence of TB infection in 1992, $P_{i(1992)}$. Therefore, the number of active TB cases reported in 1994 can be expressed as:

$$A_{i(1994)} = .04 * I_{i(1994)} + .02 * I_{i(1993)} + (.04/73) * P_{i(1992)} \qquad (2)$$

Further, if we assume that the number of new infections is directly proportional to the number of active cases, then $I_{i(1993)}$ can be expressed as follows:

$$I_{i(1993)} = I_{i(1994)} * (A_{i(1993)}/A_{i(1994)}) \qquad (3)$$

and (2) can be expressed as:

$$A_{i(1994)} = [(.02 * (A_{i(1993)}/A_{i(1994)}) + .04] * I_{i(1994)} + (.04/73) * P_{i(1992)}$$

$$A_{i(1994)} = [(.02 * R_i + .04] * I_{i(1994} + (.04/73) * P_{i(1992)} \qquad (4)$$

then solving for $I_{i(1994)}$ becomes:[2]

---

[2] Using the prevalence of TB infection in 1992 (i.e., $P_{i(1992)}$) to approximate the quantity inside the summation sign (i.e., everyone infected between 1919 and 1992 and alive in 1994) slightly overestimates the quantity inside the summation (i.e., $P_{i(1992)}$ is slightly larger than the quantity it approximates.) It includes a small number of people who were infected with TB and were alive as of 1992 and who were therefore included in the prevalence figure, but who died before 1994, and, technically, are not included in the summation. This implies that, in equation (5), a slightly larger number is being subtracted from $A_{i(1994)}$ than should be, resulting in an underestimate of the number of new infections in 1994 and an underestimate of the occupational risk.

$$I_{i(1994)} = [A_{i(1994)} - .04/73 * P_{i(1992)}]/(.02 * R_i + .04) \quad\quad (5)$$

Step 3: Estimation of $X_{i(1994)}$:

$X_{i(1994)}$, the population at risk for TB infection in state i in 1994, is estimated as follows:

$$X_{i(1994)} = N_i - P_{i(1993)} \quad\quad (6)$$

Where:

N_i is the adult population for state i as reported by U.S. Census in 1994.

$P_{i(1993)}$ is the estimated number of infected adults in state i in 1993 (i.e., prevalence of TB infection in state i among adults).

To estimate the number of adults currently at risk for TB infection in each state, the number of already infected adults (i.e., prevalence of TB infection P_i in 1993) is subtracted from the adult population in 1994.

Step 4: Estimation of population currently infected as of 1993 by state, $P_i(1993)$:

The prevalence of TB infection in each state is estimated as a function of TB infection prevalence in the U.S. in 1993 and the percent TB case rate for each state.

$$P_{i(1993)} = P_{(1993)} * (A_{i(1993)}/A_{(1993)}) \quad\quad (7)$$

Where:

$P_{(1993)}$ is the prevalence of TB infections in the U.S. in 1993 (Ex. 7-66) and

$A_{(1993)}$ is the total number of adult TB cases reported in 1993.

Estimates of TB infection prevalence in the U.S. were developed for OSHA by Dr. Christopher Murray of the Harvard Center for Population and Development Studies and are presented in Table V–6 (Ex. 7-267). The mathematical model used by Dr. Murray to estimate TB infection prevalence has been designed to capture the transmission dynamics of TB by modeling transfers between a series of age-stratified compartments using a system of differential equations. The model adjusts for various epidemiological factors known to influence the course of active TB, such as onset of infection (i.e., old vs. new infections) and the impact of immigration rates and the HIV epidemic. However, it does not differentiate among gender or race categories. The model has been successfully validated using actual epidemiological data on active TB from 1965 to 1994. The estimates of TB prevalence rates presented here are specific for adults (i.e., older than 18 years of age), which make them more appropriate for estimating risk of transmission in an occupational setting.

To estimate the number of previously infected adults in each state (P_i), the estimated national TB prevalence figure was multiplied by the active cases for each state and divided by the total number of active cases reported [see equation (7)] (i.e., the national prevalence estimate was apportioned among the states based on each state's percent contribution to active TB reported for 1993). To estimate the number

TABLE V–6
National Prevalence of TB Infection in Adults (18+)[a][b]

Year	Expected	Minimum	Maximum
1992	6.87%	6.53%	7.22%
	(12,978,461)	(12,336,150)	(13,639,663)
1993	6.64%	6.31%	6.97%
	(12,667,062)	(12,037,524)	(13,296,599)
1994	6.47%	6.14%	6.79%
	(12,449,445)	(11,814,465)	(13,065,182)

[a] Numbers in parentheses are population prevalence figures.
[b] Estimated for OSHA by Christopher Murray MD, PhD, Harvard University, Center for Population and Development Studies (Ex. 7-267).

of adults at risk of TB infection, (X_i), the number of already infected adults was subtracted from the adult population estimate for each state (see equation (6)). The number of new infections expected to have occurred in 1994 was estimated using equation (5).

The background rate of TB infection for 1994 was then estimated by dividing the number of new infections (I_i) by the number of susceptible adults in each state (X_i) (see equation (1)).

Results on estimated TB background annual infection rates for each state are presented in Table V–7(a)—Table V–7(c). In Table V–7(a) TB infection rates are based on an average value of TB infection prevalence, as estimated by Dr. Murray, in the U.S. (i.e., 12,667,062). In Table V–7(b), infection rates are based on the minimum value of TB infection prevalence in the U.S. (i.e., 12,037,524). In Table V–7(c), infection rates are based on the maximum value of TB infection prevalence in the U.S. (i.e., 13,296,599). An overall range of background annual TB infection rates was constructed by combining all three sets of infection rates and was estimated to be between 0.194 and 3.542 per 1,000 individuals at risk of TB infection, with a weighted average of 1.46 per 1,000 using state population size as weights.

Step 5 Model validation:

An alternative, but less sophisticated, way to estimate annual risk of infection, if prevalence is known in a specific age group, is to use the following formula:

$$\text{Annual Rate of Infection} = -\ln(1 - P)/d \qquad (8)$$

Where:
P is the percent prevalence of infection and
d is the average age of the population (Ex. 7-265).

In order to validate the model used by OSHA to estimate background infection rates, estimates of TB infection prevalence for 1994 were used to calculate predicted infection rates using equation (8). Based on Murray's model, TB infection prevalence

TABLE V–7(a)
Estimates of Annual Background TB Infection Rates[a]
[Referent Year 1994]

State	TB cases reported in 1994 A_i	Population size[a] N_i	Population currently infected[b] $P_{i(1993)}$	Population at risk X_i	Estimate of new infections I_i	Annual population rate of TB infection B_i
Alabama (01)	413	3,139	250,083	2,888,917	4,779	1.65
Alaska (02)	78	414	27,787	386,213	1,182	3.06
Arizona (04)	233	2,936	118,231	2,817,769	2,858	1.01
Arkansas (05)	235	1,813	107,334	1,705,666	2,906	1.70
California (06)	4,291	22,754	2,437,044	20,280,956	47,852	2.36
Colorado (08)	90	2,686	52,850	2,633,150	1,045	0.40
Connecticut (09)	144	2,487	81,182	2,405,818	1,665	0.69
Delaware (10)	51	531	26,152	504,848	671	1.33
D.C. (11)	116	451	80,092	370,908	1,162	3.13
Florida (12)	1,675	10,691	846,687	9,844,314	20,545	2.09
Georgia (13)	676	5,162	396,646	4,765,354	7,082	1.49
Hawaii (15)	234	875	132,942	742,058	25,890	3.49
Illinois (17)	1,021	8,669	622,211	8,046,789	10,994	1.37
Indiana (18)	201	4,279	129,673	4,149,327	2,083	0.50
Iowa (19)	62	2,180	31,056	2,068,943	859	0.42
Kansas (20)	77	1,864	37,049	1,826,951	1,065	0.58
Kentucky (21)	316	2,857	203,227	2,653,773	3,273	1.23
Louisiana (22)	412	3,080	185,792	2,894,208	5,582	1.93
Maine (23)	31	934	14,712	919,289	419	0.46
Maryland (24)	344	3,743	211,399	3,531,601	3,582	1.01
Massachusetts (25)	299	4,617	183,067	4,433,933	2,889	0.65
Michigan (26)	438	6,971	246,269	6,724,731	5,036	0.75
Minnesota (27)	127	3,326	68,105	3,257,895	1,413	0.43
Mississippi (28)	262	1,913	141,659	1,771,341	3,120	1.76
Missouri (29)	241	3,899	128,583	3,770,417	2,922	0.78
Montana (30)	22	618	11,987	606,013	290	0.48
Nebraska (31)	22	1,181	12,531	1,168,469	233	0.20
Nevada (32)	111	1,181	50,670	1,130,330	1,514	1.34
New Hampshire (33)	17	845	13,076	831,924	182	0.22
New Jersey (34)	764	5,973	456,579	5,516,421	8,150	1.48
New Mexico (35)	78	1,156	35,415	1,120,585	944	0.84
New York (36)	3,414	13,658	2,044,797	11,613,203	34,728	2.99
North Carolina (37)	532	5,314	298,574	5,015,426	6,000	1.20
North Dakota (38)	10	466	3,813	426,186	132	0.29
Ohio (39)	318	8,248	161,274	8,086,726	3,763	0.47
Oklahoma (40)	231	2,378	101,886	2,276,114	3,064	1.35
Oregon (41)	146	2,303	78,457	2,224,543	1,793	0.81
Pennsylvania (42)	583	9,154	379,211	8,774,789	5,886	0.67
Rhode Island (44)	47	757	31,601	725,399	495	0.68
South Carolina (45)	362	2,712	205,406	2,506,594	4,273	1.70
South Dakota (46)	26	513	8,173	504,827	342	0.68

TABLE V–7(a)—Continued
Estimates of Annual Background TB Infection Rates[a]
[Referent Year 1994]

State	TB cases reported in 1994 A_i	Population size[a] N_i	Population currently infected[b] $P_{i(1993)}$	Population at risk X_i	Estimate of new infections I_i	Annual population rate of TB infection B_i
Tennessee (47)	494	3,878	283,863	3,594,137	5,759	1.60
Texas (48)	2,276	13,077	1,199,200	11,877,800	27,306	2.30
Utah (49)	47	1,236	23,973	1,212,027	427	0.35
Vermont (50)	10	434	2,724	431,276	160	0.37
Virginia (51)	330	4,949	226,110	4,722,890	3,220	0.68
Washington (53)	241	3,935	142,729	3,792,251	2,554	0.67
West Virginia (54)	80	1,393	40,318	1,352,682	919	0.68
Wisconsin (55)	104	3,735	50,126	3,684,874	1,307	0.35
Wyoming (56)	12	339	3,814	335,186	188	0.56

[a] Expressed in thousands.
[b] Based on 6.64% rate of TB infection prevalence in the U.S. (expected)

is expected to range from 6.31% to 6.97% in 1994 among adults (18+). Using these figures and assuming the average age to be 45 years, formula (8) predicts that infection rates can range from 1.45 to 1.61 per 1,000. These results are in close agreement with OSHA's weighted average estimate of the national TB infection rate, which is 1.46 per 1,000.

Occupational Risk Estimations

OSHA used the three different data sources to obtain estimates of risk of TB infection for health care employees: the Washington State data, the North Carolina study, and the NIOSH Health Hazard Evaluation (HHE) from Jackson Memorial Hospital (Exs. 7-263, 7-7, 7-108). The Washington State data represent workplaces located in low TB prevalence areas, where TB infection control measures and engineering controls are required by state health regulations. The North Carolina data represent workplaces located in areas with moderate TB prevalence and inadequate TB infection control programs. Finally, the Jackson Memorial Hospital data are representative of county hospitals serving high-risk patients whose employees have a high frequency of exposure to infectious TB. These data sources provide information on the magnitude of the expected excess risk in three different environments, and are used to provide a range of possible values of excess risk.

Based on the Washington State data, the annual risk is expected to be 1.5 times the background rate for hospital employees, approximately 11 times the background rate for long-term care employees, 6 times the background rate for home health care workers, and double the background rate for home care employees. Based on the North Carolina data, the annual risk is expected to be approximately 5 times the

TABLE V–7(b)
Estimates of Annual Background TB Infection Rates
[Referent Year 1994[a]]

State	TB cases reported in 1994 A_i	Population size[a] N_i	Population currently infected[b] $P_{i(1993)}$	Population at risk X_i	Estimate of new infections I_i	Annual population rate of TB infection B_i
Alabama (01)	413	3,139	237,654	2,901,346	4,871	1.68
Alaska (02)	78	414	26,406	387,594	1,196	3.09
Arizona (04)	233	2,936	112,355	2,823,645	2,913	1.03
Arkansas (05)	235	1,813	102,000	1,711,000	2,967	1.73
California (06)	4,291	22,754	2,350,136	20,403,864	48,956	2.40
Colorado (08)	90	2,686	50,223	2,635,777	1,066	0.40
Connecticut (09)	144	2,487	77,147	2,409,853	1,700	0.71
Delaware (10)	51	531	24,853	506,147	681	1.34
D.C. (11)	116	451	76,111	374,889	1,192	3.18
Florida (12)	1,675	10,691	804,607	9,886,393	20,944	2.12
Georgia (13)	676	5,162	376,933	4,785,067	7,275	1.52
Hawaii (15)	234	875	126,335	748,665	2,652	3.54
Illinois (17)	1,021	8,669	591,288	8,077,712	11,260	1.39
Indiana (18)	201	4,279	123,228	4,155,772	2,136	0.51
Iowa (19)	62	2,180	29,513	2,070,487	869	0.42
Kansas (20)	77	1,864	35,208	1,828,792	1,079	0.59
Kentucky (21)	316	2,857	193,126	2,663,874	3,357	1.26
Louisiana (22)	412	3,080	176,558	2,903,442	5,667	1.95
Maine (23)	31	934	13,980	920,020	425	0.46
Maryland (24)	344	3,743	200,893	3,542,107	3,677	1.04
Massachusetts (25)	299	4,617	173,969	4,443,031	2,983	0.67
Michigan (26)	438	6,971	234,030	6,736,970	5,144	0.76
Minnesota (27)	127	3,326	64,721	3,261,279	1,448	0.44
Mississippi (28)	262	1,913	134,619	1,778,381	3,183	1.79
Missouri (29)	241	3,899	122,193	3,776,807	2,978	0.79
Montana (30)	22	618	11,391	606,609	294	0.48
Nebraska (31)	22	1,181	11,909	1,169,091	240	0.21
Nevada (32)	111	1,181	48,152	1,132,848	1,536	1.36
New Hampshire (33)	17	845	12,426	832,574	185	0.22
New Jersey (34)	764	5,973	433,887	5,539,113	8,357	1.51
New Mexico (35)	78	1,156	33,655	1,112,345	965	0.86
New York (36)	3,414	13,658	1,943,173	11,714,827	35,735	3.05
North Carolina (37)	532	5,314	283,735	5,030,265	6,138	1.22
North Dakota (38)	10	466	3,624	462,376	134	0.29
Ohio (39)	318	8,248	153,259	8,094,741	3,845	0.48
Oklahoma (40)	231	2,378	96,822	2,281,178	3,116	1.37
Oregon (41)	146	2,303	74,558	2,228,442	1,825	0.82
Pennsylvania (42)	583	9,154	360,365	8,793,635	6,047	0.69
Rhode Island (44)	47	757	30,030	726,970	506	0.70
South Carolina (45)	362	2,712	195,197	2,516,803	4,356	1.73
South Dakota (46)	26	513	7,766	505,234	350	0.69

TABLE V–7(b)—Continued
Estimates of Annual Background TB Infection Rates
[Referent Year 1994ᵃ]

State	TB cases reported in 1994 A_i	Population sizeᵃ N_i	Population currently infectedᵇ $P_{i(1993)}$	Population at risk X_i	Estimate of new infections I_i	Annual population rate of TB infection B_i
Tennessee (47)	494	3,878	269,756	3,608,244	5,875	1.63
Texas (48)	2,276	13,077	1,139,601	11,937,399	27,853	2.33
Utah (49)	47	1,236	22,782	1,213,218	446	0.37
Vermont (50)	10	434	2,589	431,411	162	0.37
Virginia (51)	330	4,949	214,873	4,734,127	3,311	0.70
Washington (53)	241	3,935	135,654	3,799,346	2,621	0.69
West Virginia (54)	80	1,393	38,315	1,354,685	941	0.69
Wisconsin (55)	104	3,735	47,634	3,687,366	1,332	0.36
Wyoming (56)	12	339	3,624	335,376	190	0.57

ᵃ Expressed in thousands.
ᵇ Based on a 6.31% rate of TB infection in the U.S.

background rate. Based on the Jackson Memorial Hospital data, the annual risk is expected to be approximately 9 times the background.

Estimates of expected excess risk of TB infection for workers with occupational exposure by state are calculated by applying the excess relative risk ratios, derived from the three occupational studies, to the overall background rate of infection for each state and are presented in table V–8(a)—table V–8(c). A range of excess risk of TB infection due to occupational exposure is constructed by using the minimum and maximum estimates of excess risk among all states for each data source. These results are presented in table V–9 and table V–10 for workers in hospitals and for workers in other work settings, respectively.

Lifetime estimates of the excess risk of TB infection were estimated based on the annual excess risk by using the formula $\{1-(1-p)^{45}\}$, where p is the annual excess risk. Lifetime excess estimates of TB infection are presented in table V–9 and table V–10. Lifetime risk estimates of developing active TB are calculated from lifetime risk estimates of TB infection assuming that, once infected, there is a 10% likelihood of progressing to active TB; these estimates are presented in table V–11 and table V–12. Further, the risk of death caused by TB is calculated from the lifetime estimates of active TB using OSHA's estimate of the TB case fatality rate (also presented in table V–11 and table V–12). The methodology used to estimate a TB case fatality rate is presented below.

As outlined in the Health Effects section, several possible outcomes are possible following an infection. Approximately 90% of all infections never progress to active disease. An estimated 10% of infections is expected to progress to active disease; most of these cases are successfully treated. However, a percentage of active TB cases develop further complications. Approximately 7.8% of active TB cases may

TABLE V–7(c)
Estimates of Annual Background TB Infection Rates
[Referent Year 1994[a]]

State	TB cases reported in 1994 A_i	Population size N_i	Population currently infected[b] $P_{i(1993)}$	Population at risk X_i	Estimate of new infections I_i	Annual population rate of TB infection B_i
Alabama (01)	413	3,139	262,512	2,876,488	4,685	1.63
Alaska (02)	78	414	29,168	384,832	1,167	3.03
Arizona (04)	233	2,936	124,107	2,811,893	2,801	1.00
Arkansas (05)	235	1,813	112,669	1,700,332	2,843	1.67
California (06)	4,291	22,754	2,595,951	20,158,049	46,720	2.32
Colorado (08)	90	2,686	55,476	2,630,524	1,024	0.39
Connecticut (09)	144	2,487	85,216	2,401,784	1,629	0.68
Delaware (10)	51	531	27,452	503,508	661	1.31
D.C.	116	451	84,072	366,928	1,131	3.08
Florida (12)	1,675	10,691	888,766	9,802,234	20,137	2.05
Georgia (13)	676	5,162	416,359	4,745,641	6,884	1.45
Hawaii (15)	234	875	139,539	735,451	2,526	3.43
Illinois (17)	1,021	8,669	653,134	8,015,866	10,721	1.34
Indiana (18)	201	4,279	136,117	4,142,883	2,029	0.49
Iowa (19)	62	2,180	32,600	2,067,401	849	0.41
Kansas (20)	77	1,864	38,891	1,825,109	1,052	0.58
Kentucky (21)	316	2,857	213,327	2,643,673	3,187	1.21
Louisiana (22)	412	3,080	195,025	2,884,975	5,496	1.91
Maine (23)	31	934	15,442	918,558	413	0.45
Maryland (24)	344	3,743	221,905	3,521,095	3,484	0.99
Massachusetts (25)	299	4,617	192,166	4,424,834	2,793	0.63
Michigan (26)	438	6,971	258,508	6,712,492	4,925	0.73
Minnesota (27)	127	3,326	71,490	3,254,510	1',377	0.42
Mississippi (28)	262	1,913	148,700	1,764,300	3,057	1.73
Missouri (29)	241	3,899	134,973	3,764,027	2,865	0.76
Montana (30)	22	618	12,582	605,418	286	0.48
Nebraska (31)	22	1,181	13,154	1,167,846	227	0.20
Nevada (32)	111	1,181	53,189	1,127,811	1,491	1.32
New Hampshire (33)	17	845	13,726	831,274	178	0.21
New Jersey (34)	764	5,973	479,270	5,493,730	7,938	1.44
New Mexico (35)	78	1,156	37,175	1,118,825	922	0.82
New York (36)	3,414	13,658	2,146,421	11,511,421	33,696	2.92
North Carolina (37)	532	5,314	313,413	5,000,587	5,859	1.17
North Dakota (38)	10	466	4,003	461,997	129	0.28
Ohio (39)	318	8,248	169,289	8,078,711	3,678	0.46
Oklahoma (40)	231	2,378	106,949	2,271,051	3,011	1.33
Oregon (41)	146	2,303	82,357	2,220,643	1,760	0.80
Pennsylvania (42)	583	9,154	398,057	8,755,943	5,722	0.66
Rhode Island (44)	47	757	33,171	723,829	483	0.67
South Carolina (45)	362	2,712	215,614	2,496,386	4,188	1.68
South Dakota (46)	26	513	8,579	504,421	334	0.67

TABLE V–7(c)—Continued
Estimates of Annual Background TB Infection Rates
[Referent Year 1994[a]]

State	TB cases reported in 1994 A_i	Population size N_i	Population currently infected[b] $P_{i(1993)}$	Population at risk X_i	Estimate of new infections I_i	Annual population rate of TB infection B_i
Tennessee (47)	494	3,878	297,971	3,580,029	5,641	1.58
Texas (48)	2,276	13,077	1,258,799	11,818,201	26,746	2.26
Utah. (49)	47	1,236	25,165	1,210,835	408	0.34
Vermont (50)	10	434	2,860	431,140	158	0.37
Virginia (51)	330	4,949	237,347	4,711,653	3,126	0.66
Washington (53)	241	3,935	149,843	3,785,157	2,485	0.66
West Virginia (54)	80	1,393	42,322	1,350,679	896	0.66
Wisconsin (55)	104	3,735	52,617	3,682,383	1,283	0.35
Wyoming (56)	12	339	4,003	334,997	185	0.55

[a] Expressed in thousands.
[b] Based on 6.97% rate of TB infection prevalence in the U.S. (maximum estimate).

take a more severe clinical course and lead to death. The TB case fatality rate was estimated using information on reported deaths caused by TB from table 8-5 of the Vital Statistics for the U.S. and cases of TB reported in CDC's TB Surveillance system for 1989 through 1991 (Exs. 7-270, 7-264). As shown in table V–13, the TB case death rate ranged from 69.94 to 89.18 per 1,000 with a 3-year average of 77.85 per 1,000 TB cases. The Agency used the 3-year average (77.85 per 1,000) for its estimate of deaths caused by TB. This estimate is in close agreement with published results from a retrospective cohort study conducted in Los Angeles County on TB cases in 1990 (Ex. 7-268). In this study, all confirmed TB cases reported in the county in 1990 were tracked and the number of deaths where TB was the direct or contributing cause was ascertained. "Contributing cause" was defined as a case of TB of such severity that it would have caused the death of the patient had the primary illness not caused death earlier. Of the 1,724 cases included in the study, TB was considered the cause of death or the contributing cause of death in 135 cases (78.31 per 1,000).

National estimates of annual and lifetime risk for TB infection, active disease and death caused by TB due to occupational exposure are computed as weighted averages of the state estimates and are presented in table V–14.

Risk Estimates for Hospital Employees
Logistic regression analysis of the Washington state hospital data indicated an increase in annual risk (47% above background) for employees with potential exposure to TB. For this particular analysis the control group was defined as those hospitals with no-known TB patients that are located in counties that did not report

TABLE V–8(a)
Occupational Risk Estimates of TB Infection
Based on the Washington State Study

State	Annual Background TB Infection Rate per 1,00 at Risk	Excess Occupational Risk	
		Annual	Lifetime
Alabama (01)	1.63–1.68	0.77–0.79	34–35
Alaska (02)	3.03–3.09	1.43–1.45	62–63
Arizona (04)	1.00–1.03	0.47–0.48	21–22
Arkansas (05)	1.67–1.73	0.79–0.81	35–36
California (06)	2.32–2.40	1.09–1.13	48–50
Colorado (08)	0.39–0.40	0.18–0.19	8–9
Connecticut (09)	0.68–0.71	0.32–0.33	14–15
Delaware (10)	1.31–1.34	0.62–0.63	27–28
District of Columbia (11)	3.08–3.18	1.45–1.49	63–65
Florida (12)	2.05–2.11	0.97–1.00	43–44
Georgia (13)	1.45–1.52	0.68–0.71	30–32
Hawaii (15)	3.43–3.54	1.61–1.66	70–72
Illinois (17)	1.34–1.39	0.63–0.66	28–29
Indiana (18)	0.50–0.51	0.23–0.24	10–11
Iowa (19)	0.41–0.42	0.19–0.20	9–9
Kansas (20)	0.58–0.59	0.27–0.28	12–12
Kentucky (21)	1.21–1.26	0.57–0.59	25–26
Louisiana (22)	1.91–1.95	0.90–0.92	39–40
Maine (23)	0.45–0.46	0.21–0.22	9–10
Maryland (24)	0.99–1.04	0.46–0.49	21–22
Massachusetts (25)	0.63–0.67	0.30–0.32	13–14
Michigan (26)	0.73–0.76	0.34–0.36	15–16
Minnesota (27)	0.42–0.44	0.20–0.21	9–9
Mississippi (28)	1.73–1.79	0.81–0.84	36–37
Missouri (29)	0.76–0.79	0.36–0.37	16–17
Montana (30)	0.47–0.48	0.22–0.23	10–10
Nebraska (31)	0.19–0.20	0.09–0.10	4–4
Nevada (32)	1.32–1.35	0.62–0.64	27–28
New Hampshire (33)	0.21–0.22	0.10–0.10	5–7
New Jersey (34)	1.44–1.51	0.68–0.71	30–31
New Mexico (35)	0.82–0.86	0.39–0.40	17–18
New York (36)	2.93–3.05	1.38–1.43	60–63
North Carolina (37)	1.17–1.22	0.55–0.57	24–25
North Dakota (38)	0.28–0.29	0.13–0.13	6–6
Ohio (39)	0.46–0.48	0.21–0.22	9–10
Oklahoma (40)	1.33–1.36	0.62–0.64	9–10
Oregon (41)	0.79–0.82	0.37–0.38	17–17
Pennsylvania (42)	0.65–0.69	0.31–0.32	14–14
Rhode Island (44)	0.67–0.70	0.31–0.33	14–15
South Carolina (45)	1.68–1.73	0.79–0.81	35–36
South Dakota (46)	0.66–0.69	0.31–0.33	14–15
Tennessee (47)	1.58–1.63	0.74–0.77	33–34

TABLE V–8(a)—Continued
Occupational Risk Estimates of TB Infection
Based on the Washington State Study

State	Annual Background TB Infection Rate per 1,00 at Risk	Excess Occupational Risk	
		Annual	Lifetime
Texas (48)	2.26–2.33	1.06–1.10	47–48
Utah (49)	0.34–0.37	0.16–0.17	7–8
Vermont (50)	0.36–0.37	0.17–0.18	8–8
Virginia (51)	0.66–0.70	0.31–0.33	14–15
Washington (53)	0.66–0.69	0.31–0.32	14–14
West Virginia (54)	0.66–0.70	0.31–0.33	14–15
Wisconsin (55)	0.35–0.36	0.16–0.17	7–8
Wyoming (56)	0.55–0.57	0.26–0.27	12–12

any active TB cases in 1994. However, an increased risk of 47% above background in the annual infection rate is expected to produce a range of 4 to 72 TB infections per 1000 exposed workers in a working lifetime, which could result in as many as 7 cases of active TB and approximately 1 death per 1,000 exposed workers.

Based on the survey of hospitals in North Carolina's western region, the expected overall risk due to occupational exposure is estimated to be 4 times the background rate. This results in an expected range of lifetime risk between 34 and 472 infections per 1,000 employees at risk for TB infection. Lifetime estimates of active TB cases resulting from these infections are expected to range between 3 and 47, resulting in as many as 4 deaths per 1,000 exposed employees at risk of TB infection. As done previously, the North Carolina study results were adjusted to reflect 1994 TB disease trends.

Based on the data from Jackson Memorial Hospital, the overall risk due to occupational exposure is estimated to be 8 times the background rate. This results in an expected range of lifetime risk between 67 and 723 infections per 1,000 employees at risk. Lifetime estimates of the number of active TB case per 100 exposed workers are expected to range between 7 and 72, resulting in as many as 6 deaths per 1,000 exposed employees at risk for TB infection.

In summary, table V–9 and table V–14 show that the annual occupational risk of infection is expected to range:

(a) From .09 to 1.66 with a weighted average of 0.68 per 1,000 for workplaces located in relatively low TB prevalence areas, and where TB infection measures and engineering controls are required;

(b) From 0.77 to 14.1 with a weighted average of 5.7 per 1,000 for workplaces located in areas with moderate TB prevalence and inadequate TB control programs; and

(c) From 1.54 to 28 with a weighted average of 11.8 per 1,000 for workplaces located in high TB prevalence areas, serving high risk patients, with high frequency of exposure to infectious TB.

TABLE V–8(b)
Occupational Risk Estimates of TB Infection
Based on the North Carolina Study

State	Annual Background TB Infection Rate per 1,00 at Risk	Excess Occupational Risk	
		Annual	Lifetime
Alabama (01)	1.63–1.68	6.48–6.68	254–260
Alaska (02)	3.03–3.09	12.07–12.28	421–427
Arizona (04)	1.00–1.03	3.96–4.11	164–169
Arkansas (05)	1.67–1.73	6.66–6.90	260–268
California (06)	2.32–2.40	9.22–9.55	341–351
Colorado (08)	0.39–0.40	1.55–1.61	67–70
Connecticut (09)	0.68–0.71	2.70–2.81	115–119
Delaware (10)	1.31–1.34	5.23–5.35	210–215
District of Columbia (11)	3.08–3.18	12.27–12.66	426–436
Florida (12)	2.05–2.11	8.18–8.43	309–317
Georgia (13)	1.45–1.52	5.77–6.05	229–239
Hawaii (15)	3.43–3.54	13.67–14.10	462–472
Illinois (17)	1.34–1.39	5.32–5.55	214–221
Indiana (18)	0.50–0.51	1.95–2.05	84–88
Iowa (19)	0.41–0.42	1.64–1.67	71–73
Kansas (20)	0.58–0.59	2.29–2.35	98–100
Kentucky (21)	1.21–1.26	4.80–5.02	195–202
Louisiana (22)	1.91–1.95	7.58–7.77	290–296
Maine (23)	0.45–0.46	1.79–1.84	77–80
Maryland (24)	0.99–1.04	3.94–4.13	163–170
Massachusetts (25)	0.63–0.67	2.51–2.67	107–113
Michigan (26)	0.73–0.76	2.92–3.04	123–128
Minnesota (27)	0.42–0.44	1.68–1.77	73–77
Mississippi (28)	1.73–1.79	6.90–7.12	268–275
Missouri (29)	0.76–079	3.03–3.14	128–132
Montana (30)	0.47–0.48	1.88–1.93	81–83
Nebraska (31)	0.19–0.20	0.77–0.82	34–36
Nevada (32)	1.32–1.35	5.26–5.40	211–216
New Hampshire (33)	0.21–0.22	0.85–0.88	38–39
New Jersey (34)	1.44–1.51	5.75–6.01	229–237
New Mexico (35)	0.82–0.86	3.28–3.42	137–143
New York (36)	2.93–3.05	11.65–12.14	410–423
North Carolina (37)	1.17–1.22	4.66–4.86	190–196
North Dakota (38)	0.28–0.29	1.11–1.16	49–50
Ohio (39)	0.46–0.48	1.81–1.89	78–82
Oklahoma (40)	1.33–1.36	5.28–5.44	212–216
Oregon (41)	0.79–0.82	3.15–3.26	133–137
Pennsylvania (42)	0.65–0.69	2.60–2.74	111–116
Rhode Island (44)	0.67–0.70	2.66–2.77	113–117
South Carolina (45)	1.68–1.73	6.68–6.89	260–267
South Dakota (46)	0.66–0.69	2.64–2.76	112–117
Tennessee (47)	1.48–1.63	6.26–6.48	247–254

TABLE V–8(b)—Continued
Occupational Risk Estimates of TB Infection
Based on the North Carolina Study

State	Annual Background TB Infection Rate per 1,00 at Risk	Excess Occupational Risk	
		Annual	Lifetime
Texas (48)	2.26–2.33	9.01–9.29	334–343
Utah (49)	0.34–0.37	1.34–1.46	59–64
Vermont (50)	0.36–0.37	1.46–1.49	63–65
Virginia (51)	0.66–0.70	2.64–2.68	112–118
Washington (53)	0.66–0.69	2.61–2.75	111–116
West Virginia (54)	0.66–0.70	2.64–2.77	112–117
Wisconsin (55)	0.35–0.36	1.39–1.44	61–73
Wyoming (56)	0.55–0.57	2.20–2.25	94–97

Similarly, the lifetime occupational risk is expected to range:

(a) From 4 to 72 with a weighted average of 30 per 1,000 for workplaces located in relatively low TB prevalence areas, and where TB infection measures and engineering controls are required;

(b) From 34 to 472 with a weighted average of 219 per 1,000 for workplaces located in areas with moderate TB prevalence and inadequate TB control programs; and

(c) From 67 to 723 with a weighted average of 386 per 1,000 for workplaces located in high TB prevalence areas, serving high risk patients, with high frequency of exposure to infectious TB.

Risk estimates derived from either study (Washington State or North Carolina) represent an overall rate of occupational risk, because both studies include PPD skin testing results from the entire hospital employee population, whereas the Jackson Memorial study addresses the occupational risk to workers where exposure to infectious TB is highly probable.

Although the exact compliance rate is not known, hospitals in Washington State have been required to implement the CDC TB guidelines with respect to engineering controls (requiring isolation rooms with negative pressure) and infection control measures (advocating early patient identification, employee training, respiratory protection, and PPD testing).

Neither the facilities in North Carolina nor Jackson Memorial had engineering controls fully implemented at the time these data were collected. Early identification of suspect TB patients has always been recommended in North Carolina. However, engineering controls in isolation rooms were either not present or did not function properly because of modifications in the physical structure of the building (i.e., isolation rooms had been subdivided using partitions, air ducts had been re-directed because of remodeling, etc.). Tuberculin skin testing was very inconsistent and sporadic. In addition, employee training and use of respiratory protection were not emphasized.

TABLE V–8(c)
Occupational Risk Estimates of TB Infection
Based on the Jackson Memorial Hospital Study

State	Annual Background TB Infection Rate per 1,00 at Risk	Excess Occupational Risk	
		Annual	Lifetime
Alabama (01)	1.63–1.68	13.33–13.75	454–464
Alaska (02)	3.03–3.09	24.84–25.27	678–684
Arizona (04)	1.00–1.03	8.16–8.45	308–317
Arkansas (05)	1.67–1.73	13.69–14.20	462–475
California (06)	2.32–2.40	18.98–19.66	568–591
Colorado (08)	0.39–0.40	3.19–3.31	134–139
Connecticut (09)	0.68–0.07	5.55–5.78	222–230
Delaware (10)	1.31–1.34	10.75–11.01	385–392
District of Columbia (11)	3.08–3.18	25.24–26.04	683–695
Florida (12)	2.05–2.11	16.83–17.35	534–545
Georgia (13)	1.45–1.52	11.88–12.45	416–431
Hawaii (15)	3.43–3.54	28.13–29.01	723–734
Illinois (17)	1.34–1.39	10.95–11.42	391–404
Indiana (18)	0.50–0.51	4.01–4.21	165–173
Iowa (19)	0.41–0.42	3.36–3.44	141–144
Kansas (20)	0.58–0.59	4.72–4.83	192–196
Kentucky (21)	1.21–1.26	9.87–10.32	360–373
Louisiana (22)	1.91–1.95	15.60–15.99	507–516
Maine (23)	0.45–0.46	3.69–3.79	153–157
Maryland (24)	0.99–1.02	8.11–8.50	307–319
Massachusetts (25)	0.63–0.67	5.17–5.50	208–220
Michigan (26)	0.73–0.76	6.01–6.25	238–246
Minnesota (27)	0.42–0.44	3.46–3.64	145–151
Mississippi (28)	1.73–1.79	14.19–14.66	474–485
Missouri (29)	0.76–0.79	6.23–6.46	245–253
Montana (30)	0.47–0.58	3.87–3.96	160–164
Nebraska (31)	0.19–0.20	1.59–1.68	69–73
Nevada (32)	1.32–1.35	10.83–11.10	387–395
New Hampshire (33)	0.21–0.22	1.76–1.82	76–79
New Jersey (34)	1.44–1.51	11.83–12.36	415–429
New Mexico (35)	0.82–0.86	6.75–7.05	263–273
New York (36)	2.93–3.05	23.96–24.98	664–680
North Carolina (37)	1.17–1.22	9.60–9.99	352–364
North Dakota (38)	0.28–0.29	2.29–2.38	98–102
Ohio (39)	0.46–0.48	3.73–3.89	155–161
Oklahoma (40)	1.33–1.36	10.86–11.19	388–397
Oregon (41)	0.79–0.82	6.49–6.71	254–261
Pennsylvania (42)	0.65–0.69	5.35–5.63	214–224
Rhode Island (44)	0.67–0.70	5.47–5.70	218–227
South Carolina (45)	1.68–1.73	13.74–14.18	463–474
South Dakota (46)	0.66–0.69	5.42–5.68	216–226
Tennessee (47)	1.58–1.63	12.91–13.33	443–453

TABLE V–8(c)—Continued
Occupational Risk Estimates of TB Infection
Based on the Jackson Memorial Hospital Study

State	Annual Background TB Infection Rate per 1,00 at Risk	Excess Occupational Risk	
		Annual	Lifetime
Texas (48)	2.26–2.33	18.54–19.10	569–580
Utah (49)	0.34–0.37	2.76–3.01	117–127
Vermont (50)	0.36–0.37	2.99–3.07	126–129
Virginia (51)	0.66–0.70	5.43–5.73	217–228
Washington (53)	0.66–0.69	5.38–5.65	215–225
West Virginia (54)	0.66–0.70	5.44–5.70	217–226
Wisconsin (55)	0.35–0.36	2.86–2.96	121–125
Wyoming (56)	0.55–0.57	4.53–4.64	185–189

TABLE V–9
Occupational Risk Estimates for Hospital Employees[a]

Source	Overall risk/ (exposed)	Background risk based on study	Excess risk based on study (percent)	Range of excess occupational risk[c]	
				Annual	Lifetime
Washington State 1994 data	1.24/1000	0.88/1000	47	0.09–1.66	4.1–72.2
North Carolina Western Counties	[b]5.98/1000	[d]1.20/1000	398	0.77–14.1	34.2–472
Jackson Memorial (1991)	31.7/1000	3.5/1000	795	1.54–28.2	67.1–723

[a] Background TB infection rate ranges from 0.194 to 3.542 per 1,000 at risk for TB infection.

[b] Adjusted for 1994, i.e., 5.98=7.2*(532/641)

[c] The range reflects regional differences in TB prevalence as well as inherent uncertainty in the estimate of TB infection prevalence in the U.S., as estimated by Dr. Christopher Murray, and used in the internal calculations of annual background TB infection rate.

[d] State-wide estimate of population risk for North Carolina, shown in Table V–3(a).

By 1991, Jackson Memorial had most of the engineering controls in place in the HIV ward (where the first outbreak took place) and in selected areas with high TB exposure, but not in the entire hospital. However, the staff training program was still being developed and respiratory protection was not always adequate. Although exposures had been greatly reduced, "high risk" procedures were still being performed in certain areas of the hospital without adequate engineering controls, such as the Special Immunology clinic where HIV-TB patients received pentamidine treatments. Like the hospitals in the North Carolina study, Jackson Memorial represents a working environment that serves a patient population known to have high

TABLE V–10
Occupational Risk Estimates for Other Work Settings[a,b]

Type	Overall risk/ (exposed)	Background risk State-wide[c]	Excess risk based on study (percent)	Range of excess occupational risk[d] Annual	Range of excess occupational risk[d] Lifetime
Long-term Care	9.8/1000	0.8756/1000	1019	1.98–36.1	85–807
Home Health Care	5.06/1000	0.8756/1000	478	0.93–16.9	40.9–526
Home Care	1.86/1000	0.8756/1000	112	0.22–3.97	9.7–164

[a] Background TB infection rate ranges from 0.194 to 3.542 per 1,000 employees at risk of infection.
[b] Based on the Washington State data.
[c] Background rate for this analysis is assumed to be the same as in the case-control analysis of the Washington State hospital data (i.e. 0.8756 per 1,000 employees).
[d] The range reflects regional differences in TB prevalence as well as inherent uncertainty in the estimate of TB infection prevalence in the U.S., as estimated by Dr. Christopher Murray, and used in the internal calculations of annual background TB infection rate.

TABLE V–11
Lifetime Occupational Risk Estimates for Hospital Employees[a,b,c]

Source	TB infection[d]	Active disease[e]	Death caused by TB
Washington State (1994)	4.1–72.2	0.4–7.2	0.03–0.6
North Carolina Western Region	34.2–472	3.4–47.2	0.3–3.7
Jackson Memorial Hospital (Miami)	67.1–723	6.7–72.3	0.5–5.6

[a] Risk estimates reflect excess risk due to occupational exposure and are expressed per 1,000 employees at risk.
[b] Estimates of death caused by TB due to occupational exposure are derived based on an estimated TB case death rate of 77.85 per 1,000 TB cases and are estimated by multiplying the lifetime active disease rate by .07785.
[c] The ranges of risk presented in this TABLE reflect expected variance in the annual background TB infection rate by state. They are estimated based on the assumption that the annual background TB infection rate ranges from 0.194 to 1.542 per 1,000 employees at risk.
[d] Lifetime infection rate is estimated by $(1-(1-p)^{45})$, where p is the annual excess TB infection rate due to occupational exposure.
[e] Lifetime active disease rate is estimated to be 10% of lifetime infection rate.

TB prevalence. In addition, Jackson Memorial only tested employees with patient contact in areas where active TB had been detected.

Risk Estimates for Workers in Other Work Settings
In long-term care facilities for the elderly there is also a significantly increased likelihood that employees will encounter individuals with infectious TB. Persons

TABLE V–12
Lifetime Occupational Risk Estimates for Employees in Other Work Settings[a,b,c]

Work setting	TB infection[d]	Active disease[e]	Death caused by TB
Long-term Care	85–807	8.5–80.7	0.7–6.2
Home Health Care	40.9–536	4.1–53.6	0.3–4.2
Home Care	9.7–164	1.0–16.4	0.1–1.3

[a] Risk estimates reflect excess risk due to occupational exposure and are expressed per 1,000 employees at risk of TB infection.

[b] Estimates of death caused by TB due to occupational exposure are derived based on an estimated TB case death rate of 77.85 per 1,000 cases and are estimated by multiplying the lifetime active disease rate by .07785.

[c] The ranges of risk presented in this TABLE reflect expected variance in the annual background TB infection rate by state. They are estimated based on the assumption that the annual background TB infection rate ranges from 0.194 to 3.542 per 1,000 employees at risk.

[d] Lifetime infection rate is estimated by $(1-(1-p)^{45})$, where p is the annual excess TB infection rate due to occupational exposure.

[e] Lifetime active disease rate is estimated to be 10% of lifetime infection rate.

TABLE V–13
TB Case Death Rates for Adults (18+)

Year	Number of deaths[a]	Number of TB cases[b]	TB case death rate[c]
1991	1,700	24,307	69.94
1990	1,796	23,795	75.48
1989	1,956	21,934	89.18
3-year Average	1,817	23,345	77.85

[a] Source: Vital Statistics for the U.S., Table 8–5, (age 20+).

[b] Source: CDC, TB surveillance system, (age 18+).

[c] Rate expressed per 1,000 TB cases. Any deaths caused by TB in persons 18 or 19 years of age are not included in the numerator.

over the age of 65 constitute a large proportion of the TB cases in the United States. In 1987, CDC reported that persons aged 65 and over accounted for 27% (6150) of the reported cases of active TB in the U.S., although they account for only 12% of the U.S. population. Many of these individuals were infected in the past and advancing age and decreasing immunocompetence have caused them to develop active disease. In 1990 the CDC estimated that approximately 10 million people were infected with TB. As the U.S. population steadily ages, many of these latent infections may progress to active disease. Because elderly persons represent a large proportion of the nation's nursing home residents and because the elderly represent a large proportion of the active cases of TB, there is an increased likelihood that

TABLE V–14
Average Occupational Risk Estimates[a,b] per 1,000 Workers at Risk

Work setting	Annual TB infection	Lifetime TB infection	Lifetime active TB	Death caused by TB[c]
Hospitals:				
WA	0.68	30	3.0	0.2
NC	5.7	219	22.0	1.7
JM	11.8	386	38.6	3.0
Long-term Care	14.6	448	44.8	3.5
Home Health Care	6.9	225	25.5	2.0
Home Care	1.6	69	6.9	0.5

[a] Weighted by each state's population in 1994.

[b] Risk estimates reflect excess risk due to occupational exposure and are expressed per 1,000 employees at risk.

[c] Number of deaths caused by TB due to occupational exposure are derived based on an estimated TB case death rate of 77.85 per 1,000 cases and are computed by multiplying the lifetime active disease rate by .07785.

employees at long-term care facilities for the elderly will encounter individuals with infectious TB.

Similarly, there are other occupational settings that serve high-risk client populations and thus have an increased likelihood of encountering individuals with infectious TB. For example, hospices, emergency medical services, and home-health care services provide services to client populations similar to those in hospitals and thus are likely to experience similar risks.

OSHA used information from the 1994 Washington state PPD skin testing survey to estimate occupational risk for workers in long-term care, home health care, and home care. Annual estimates of excess risk for TB infection are presented in TABLE V–10 and lifetime estimates for TB infection, active TB, and death caused by occupational TB are presented in TABLE V–12.

Based on the Washington State data, the overall annual excess risk for TB infection is estimated to be 10-fold over background for workers in long-term care. This results in an expected range of lifetime risk of between 85 and 800 infections per 1,000 employees at risk for TB infection. Lifetime estimates of the number of active TB cases resulting from these infections range from 9 to 81 and are projected to cause as many as 6 deaths per 1,000 exposed employees at risk of TB infection. Similarly, the overall annual excess risk of TB infection for workers in home health care is estimated to be approximately 500% above background. This results in an expected range of lifetime risk of between 41 and 536 infections per 1,000 employees at risk for TB infection. Lifetime estimates of the number of active TB cases range from 4 to 54 per 1,000, and are projected to cause as many as 4 deaths per 1,000 exposed employees at risk of TB infection. Similarly, the overall annual excess risk of TB infection for workers in home care is estimated to be approximately 100% above background. This results in an expected range of lifetime risk of between

10 and 164 infections per 1,000 employees at risk for TB infection. Lifetime estimates of the number of active TB cases range from 1 to 16, and are expected to result in approximately 1 death per 1,000 exposed employees at risk of TB infection.

Clearly, employees in all three groups (long-term care for the elderly, home health care, and home care) have higher risks than hospital employees in Washington. This could be attributed, in part, to the lack of engineering controls in these work settings. That respirators may be used only intermittently may also play a role. Although workers in these three groups are encouraged by local health authorities to use respiratory protection while tending to a suspect TB patient, the actual rate of respirator usage is difficult to ascertain. A third factor that may contribute to higher risk in these work settings is delayed identification of suspect TB patients due to confounding symptoms presented by the individuals. For example, many long-term care residents exhibit symptoms of persistent coughing from decades of smoking. Consequently, an individual in long-term care with a persistent cough may be infectious for several days before he or she is identified as having suspected infectious TB.

QUALITATIVE ASSESSMENT OF RISK FOR OTHER OCCUPATIONAL SETTINGS

The quantitative estimates of the risk of TB infection discussed above are based primarily upon data from hospitals and selected other health care settings. Data from hospitals and certain health care settings were selected because OSHA believes that these data represent the best information available to the Agency for purposes of quantifying the occupational risks of TB infection and disease. However, as discussed above, it is their exposure to aerosolized *M. tuberculosis* that places these workers at risk of infection and not factors unique to these particular kinds of health care activities. Thus, OSHA believes that the risk estimates derived from hospitals and selected other work settings can be used to describe the potential range of risks for other health care and other occupational settings in which workers can reasonably anticipate frequent and substantial exposure to aerosolized *M. tuberculosis*.

In order to extrapolate the quantitative risk estimates calculated for hospital employees and other selected health care settings, OSHA, as a first step, identified risk factors that place employees at risk of exposure. Some amount of exposure to TB could occur in any workplace in the United States. TB is an infectious disease that occurs in the community and thus, individuals may bring the disease into their own workplace or to other businesses or work settings that they may visit. However, there are particular kinds of work settings where risk factors are present that substantially increase the likelihood that employees will be frequently exposed to aerosolized *M. tuberculosis*. First among these factors is the increased likelihood of exposure to individuals with active, infectious TB. Individuals who are infected with TB have a higher risk of developing active TB if they are (1) immunocompromised (e.g., elderly, undergoing chemotherapy, HIV positive), (2) intravenous drug users, or (3) medically underserved and of generally poor health status (Exs. 6-93 and 7-50). Thus, in work settings in which the client population is composed of a high proportion of individuals who are infected with TB, are immunocompromised, are intravenous drug users or are of poor general health status, there is a greatly increased likelihood that employees will routinely encounter individuals with infectious TB and be exposed to aerosolized *M. tuberculosis*. A second factor that places employees at high risk of exposure to aerosolized *M. tuberculosis* is the performance of high-hazard procedures, i.e., procedures performed on individuals with suspected or confirmed infectious TB where there is a high likelihood of the generation of droplet nuclei. A third factor that places employees at risk of exposure is the environmental conditions at the work setting. Work settings that have overcrowded conditions or poor ventilation will facilitate the transmission of TB. Thus, given that a case of infectious TB does occur, the conditions at the work setting itself may promote the transmission of disease to employees who share airspace with the individual(s) with infectious TB.

The second step in extrapolating the quantitative risks is to identify the types of work settings which have some or all of the risk factors outlined above. Once these work settings have been identified, OSHA believes that it is reasonable to assume that the quantitative risk estimates calculated for hospitals and other selected health care settings can be used to describe the risks in the identified work settings.

CORRECTIONAL FACILITIES

Employees in correctional facilities or other facilities that house inmates or detainees have an increased likelihood of frequent exposure to individuals with infectious TB. Many correctional facilities have a higher incidence of TB cases in comparison to the incidence in the general population. In 1985, the CDC estimated that the incidence of TB among inmates of correctional facilities was more than three times higher than that for nonincarcerated adults aged 15-64 (Ex. 3-33). In particular, in states such as New Jersey, New York, and California, the increased incidence of annual TB cases in correctional facilities ranged from 6 to 11 times greater than that of the general population for their respective states (Exs. 7-80 and 3-33). A major factor in the increased incidence of TB cases in correctional facilities is the fact that the population of correctional facilities is over-represented by individuals who are at greater risk of developing active disease, e.g., persons from poor and minority groups who may suffer from poor nutritional status and poor health care, intravenous drug users, and persons infected with HIV. Similarly, certain types of correctional facilities, such as holding facilities associated with the Immigration and Naturalization Service, may have inmates/detainees from countries with a high incidence of TB. For foreign-born persons arriving in the U.S., the case rate of TB in 1989 was estimated to be 124 per 100,000, compared to an overall TB case rate of 9.5 per 100,000 for the U.S. (Ex. 6-26). Moreover, in the period from 1986 to 1989, 22% of all reported cases of TB disease occurred in the foreign-born population. Given the increased prevalence of individuals at risk for developing active TB, there is an increased likelihood that employees working in these facilities will encounter individuals with infectious TB. In addition, environmental factors such as overcrowding and poor ventilation facilitate the transmission of TB. Thus, given that a case of infectious TB does occur, the conditions in the facility itself promote the transmission of the disease to other inmates and employees in the facility who share airspace.

As discussed in the Health Effects section, a number of outbreak investigations (Exs. 6-5, 6-6) have shown that where there has been exposure to aerosolized *M. tuberculosis* in correctional facilities, the failure to promptly identify individuals with infectious TB and provide appropriate infection control measures has resulted in employees being infected with TB. These studies demonstrate that, as in hospitals or health care settings, where there is exposure to aerosolized TB bacilli and where effective control measures are not implemented, exposed employees are at risk of infection. Thus, estimates based on the risk observed among employees in hospitals and in selected other work settings that involve an increased likelihood of exposure can be appropriately applied to employees in correctional facilities.

Recently, scientists at NIOSH have completed a prospective study of the incidence of TB infection among New York State correctional facilities employees (Ex. 7-288). This study is the first prospective study of TB infection among employees in correctional facilities in an entire state. Other studies have reported on contact investigations, which seek to identify recent close contacts with an index case and determine who might subsequently have been infected. Studies based on contact investigations have the advantage of a good definition of potential for exposure and they serve to identify infected persons for public health purposes. On the other hand, prospective studies of an entire working group have the advantage of covering the entire population potentially at risk, of considering all inmate cases simultaneously as potential sources of infection, and, most importantly, of permitting the calculation of incidence rates and risk attributable to occupational exposure.

Following an outbreak of active TB among inmates that resulted in transmission to employees in 1991, the state of New York instituted a mandatory annual tuberculin skin testing program to detect TB infection among employees. The authors used data from the first two years of testing to estimate the incidence of TB infection among 24,487 employees of the NY Department of Corrections. Subjects included in the study had to have two sequential PPD skin tests, have a negative test the first year, and have complete demographic information. The overall conversion rate was estimated to be 1.9%. Preliminary results show that after controlling for age, ethnicity, gender, and residence in New York City, corrections offices and medical personnel, working in prisons with inmate active TB cases, had odds ratios of TB infection of 1.64 and 2.39, respectively, compared to maintenance and clerical personnel who had little opportunity for prisoner contact. Based on these results, the annual excess risk due to occupational exposure is estimated to be 1.22% and 2.64% for corrections officers and medical personnel, respectively. This translates into lifetime occupational risks of 423 and 700 per 1,000 exposed employees, respectively. In prisons with no known inmate TB cases, there were no significant differences in TB infection rates among employees in different job categories.

HOMELESS SHELTERS

Employees in homeless shelters also have a significantly increased likelihood of frequent exposure. A high prevalence of TB infection and disease is common in many homeless shelters. Screening in selected shelters has shown the prevalence of TB infection to range from 18 to 51% (Ex. 6-15). Many shelter residents also possess characteristics that impair their immunity and thus place them at greater risk of developing active disease. For example, homeless persons often suffer from poor nutrition and poor overall health status, and they also have poor access to health care. In addition, they may suffer from alcoholism, drug abuse and infection with HIV. Screening of selected shelters has shown the prevalence of active TB disease to range from 1.6 to 6.8% (Ex. 6-15). Thus, there is an increased likelihood that employees at homeless shelters will frequently encounter individuals with infectious TB in the course of their work.

In addition, as in the case for correctional facilities, homeless shelters also tend to be overcrowded and have poor ventilation, factors that promote the transmission of disease and place shelter residents and employees at risk of infection. Outbreaks reported among homeless shelters (Exs. 7-51, 7-75, 7-73, 6-25) also provide evidence that where there is exposure to individuals with infectious TB and effective infection control measures are not implemented, employees are at risk of infection. It is reasonable to assume, therefore, that risk estimates calculated for hospital employees who have an increased likelihood of exposure to individuals with infectious TB can be used to estimate the risks for homeless shelter employees.

FACILITIES THAT PROVIDE TREATMENT FOR DRUG ABUSE

Employees in facilities that provide treatment for drug abuse have an increased likelihood of frequent exposure to individuals with infectious TB. Surveys of selected U.S. cities by the CDC have shown the prevalence of TB infection among the clients of drug treatment centers to range from approximately 10% to 13% (Ex. 6-8). Clients of these centers are also generally at higher risk of developing active disease. The clients typically come from medically underserved populations and may suffer from poor overall health status. As discussed in the Health Effects section, drug dependence has also been shown to be a possible risk factor in the development of active TB. Moreover, many of the drug treatment center clients are intravenous drug users and are infected with HIV, placing these individuals at an increased risk of developing active TB. Given these risk factors for the clients served at drug treatment centers, there is an increased likelihood that employees in these work settings will be exposed frequently to individuals with infectious TB.

MEDICAL LABORATORIES

Medical laboratory work is a recognized source of occupational hazards. CDC considers workers in medical laboratories that handle *M. tuberculosis* to be at high risk for occupational transmission of TB either because of the volume of material handled by routine diagnostic laboratories or the high concentrations of pathogenic agents often handled in research laboratories.

Few surveys of laboratory-acquired infections have been undertaken; most reports are of small outbreaks in specific laboratories. Sulkin and Pike's study of 5,000 laboratories suggested that brucellosis, tuberculosis, hepatitis, and enteric diseases are among the most common laboratory-acquired infections (Ex. 7-289). In 1957, Reid noted that British medical laboratory workers had a risk of acquiring tuberculosis two to nine times that of the general population (Ex. 7-289). This result was validated in 1971 by Harrington and Channon in their study of medical laboratories (Ex. 7-289). A retrospective postal survey of approximately 21,000 medical laboratory workers in England and Wales showed a five-times increased risk of developing active TB among these workers as compared with the general population. Technicians were at greater risk, especially if they worked in anatomy departments. A similar survey carried out in 1973 of 3,000 Scottish medical laboratory workers corroborates the results from England and Wales. Three cases, one doctor and two technicians, were noted in the 1973 survey, which resulted in an overall incidence rate of 109 per 100,000 person-years. The general population incidence rate for active TB was 26 per 100,000 person-years, giving a risk ratio of 4.2 (Ex. 7-289).

The studies reviewed in this section indicate that workers in medical laboratories with potential for exposure to *M. tuberculosis* during the course of their work have a several-fold (ranging from 2- to 9-fold) increased risk of developing active disease compared with the risk to the general population. Although these studies were conducted over two decades ago, they represent the most recent data available to the Agency, and OSHA has no reason to believe that the conditions giving rise to the risk of infection at that time have changed substantially in the interim. The Agency is not aware of any more current data on transmission rates in medical laboratories. OSHA solicits information on additional studies addressing occupational exposure to active TB in laboratories; such studies would then be considered by OSHA in the development of a final rule.

OTHER WORK SETTINGS AND ACTIVITIES

In addition to the information available for correctional facilities, homeless shelters, and facilities that provide treatment for drug abuse, there are other work settings and activities where there is an increased likelihood of frequent exposure to aerosolized *M. tuberculosis*. For example, hospices serve client populations similar to those of hospitals and perform similar services for these individuals. Individuals who receive care in hospices are likely to suffer from medical conditions (e.g., HIV disease, end-stage renal disease, certain cancers) that increase their likelihood of developing active TB disease once infected. Thus, employees providing hospice care have an increased likelihood of being exposed to aerosolized *M. tuberculosis*. CDC has recommended that hospices follow the same guidelines for controlling TB that hospitals follow.

Emergency medical service employees also have an increased likelihood of encountering individuals with infectious TB. Like hospices, emergency medical services cater to the same high risk client populations as hospitals. Moreover, emergency medical services are often used to transport individuals identified with suspected or confirmed infectious TB from various types of health care settings to facilities with isolation capabilities.

In addition, other types of services (e.g., social services, legal counsel, education) are provided to individuals who have been identified as having suspected or confirmed infectious TB and have been placed in isolation or confined to their homes. Employees who provide social welfare services, teaching, law enforcement or legal services to those individuals who are in AFB isolation are exposed to aerosolized *M. tuberculosis*. In particular, employees performing high-hazard procedures are likely to generate aerosolized *M. tuberculosis* by virtue of the procedure itself. Thus, employees providing these types of services also have an increased likelihood of exposure to aerosolized *M. tuberculosis* and are therefore likely to experience risks similar to those described above for hospital workers.

Although they do not have contact with individuals with infectious TB, employees who repair and maintain ventilation systems which carry air contaminated with *M. tuberculosis* and employees in laboratories who manipulate tissue samples or cultures contaminated with *M. tuberculosis* also have an increased likelihood of being exposed to aerosolized *M. tuberculosis*. Like employees in the work settings discussed above, these employees have an increased risk of frequent exposure to aerosolized *M. tuberculosis*.

Therefore, OSHA believes that the quantitative risk estimates derived from data observed among health care workers in the hospital setting can be generally used to describe the potential range of risks for workers in other occupational settings where there is a reasonable anticipation of exposure to aerosolized *M. tuberculosis*. The reasonableness of this assumption is supported by the overall weight of evidence of the available health data. As discussed in the Health Effects section, epidemiological studies, case reports and outbreak investigations have shown that in correctional facilities, homeless shelters, long-term care facilities for the elderly, drug treatment centers, and laboratories where appropriate TB infection control programs have not been implemented, employees have become infected with TB as a result of occupational exposure to individuals with infectious TB or to other sources of aerosolized *M. tuberculosis*. Thus, although the data on employee conversion rates in other work settings cannot be used to directly quantify the occupational risk of infection for those work settings, there is strong evidence that employees in various work settings other than hospitals can reasonably be anticipated to have exposure to aerosolized *M. tuberculosis* and that TB can be transmitted in these workplaces when appropriate TB infection control programs are not implemented.

SIGNIFICANCE OF RISK

Section 6(b)(5) of the OSH Act vests authority in the Secretary of Labor to issue health standards. This section provides, in part, that:

> The Secretary, in promulgating standards dealing with toxic materials or harmful physical agents under this subsection, shall set the standard which most adequately assures, to the extent feasible, on the basis of the best available evidence, that no employee will suffer impairment of health or functional capacity even if such employee has regular exposure to the hazard dealt with by such standard for the period of his working life.

OSHA's overall analytical approach to making a determination that workplace exposure to certain hazardous conditions presents a significant risk of material impairment of health is a four step process consistent with interpretations of the

OSH Act and rational, objective policy formulation. In the first step, a quantitative risk assessment is performed where possible and considered with other relevant information to determine whether the substance to be regulated poses a significant risk to workers. In the second step, OSHA considers which, if any, of the regulatory alternatives being considered will substantially reduce the risk. In the third step, OSHA examines the body of "best available evidence" on the effects of the substance to be regulated to set the most protective requirements that are both technologically and economically feasible. In the fourth and final step, OSHA considers the most cost-effective way to achieve the objective.

In the Benzene decision, the Supreme Court indicated when a reasonable person might consider the risk significant and take steps to decrease it. The Court stated:

> It is the Agency's responsibility to determine in the first instance what it considers to be "significant" risk. Some risks are plainly acceptable and others are plainly unacceptable. If, for example, the odds are one in a billion that a person will die from cancer by taking a drink of chlorinated water, the risk could not be considered significant. On the other hand, if the odds are one in a thousand that regular inhalation of gasoline vapors that are 2% benzene will be fatal, a reasonable person might well consider the risk significant and take the appropriate steps to decrease or eliminate it. (*I.U.D.* v. *A.P.I.*), 448 U.S. at 655).

The Court indicated that "while the Agency must support its findings that a certain level of risk exists with substantial evidence, we recognize that its determination that a particular level of risk is 'significant' will be based largely on policy considerations." The Court added that the significant risk determination required by the OSH Act is "not a mathematical straitjacket" and that "OSHA is not required to support its findings with anything approaching scientific certainty." The Court ruled that "a reviewing court (is) to give OSHA some leeway where its findings must be made on the frontiers of scientific knowledge and that the Agency is free to use conservative assumptions in interpreting the data with respect to carcinogens, risking error on the side of overprotection rather than underprotection." (448 U.S. at 655, 656).

As a part of the overall significant risk determination, OSHA considers a number of factors. These include the type of risk presented, the quality of the underlying data, the reasonableness of the risk assessments, and the statistical significance of the findings.

The hazards presented by the transmission of tuberculosis, such as infection, active disease, and death are very serious, as detailed above in the section on health effects. If untreated, 40-60% of TB cases have been estimated to result in death (Exs. 5-80, 7-50, 7-66). Fortunately, TB is a treatable disease. The introduction of antibiotic drugs for TB has helped to reduce the mortality rate by 94% since 1953 (Ex. 5-80). However, TB is still a fatal disease in some cases. From 1989-1991 CDC reported 5,452 deaths among adults from TB (see TABLE V-13, Risk Assessment section). In addition, there has been an increase in certain forms of drug-resistant TB, such as MDR-TB, in which the tuberculosis bacilli are resistant to one or more of the front line drugs such as isoniazid and rifampin, two of the most effective anti-TB drugs.

The information available today is not adequate to estimate the future course of MDR-TB, but the reduction in the potential of transmitting this deadly form of the disease is itself another benefit of this standard. The current data indicate that among MDR-TB cases, the risk of death is increased compared to drug-susceptible forms of the disease. A CDC investigation of 8 outbreaks of MDR-TB revealed that among 253 people infected with MDR-TB, 75% died within a period 4 to 16 weeks after the time of diagnosis (Ex. 38-A). MDR-TB may be treated, but due to the difficulty in finding adequate therapy which will control the bacilli's growth, individuals with this form of the disease may remain infectious for longer periods of time, requiring longer periods of hospitalization, additional lost worktime, and an increased likelihood of spreading TB infection to others until treatment renders the patient non-infectious. Because of the difficulty in controlling these drug-resistant forms of the disease with antibiotics, progressive lung destruction may progress to the point where it is necessary to remove portions of the lung to treat the advance of the disease.

The OSH Act directs the Agency to set standards that will adequately assure, to the extent feasible, that no employee will suffer "material impairment of health or functional capacity." TB infection represents a material impairment of health that may lead to active disease, tissue and organ damage, and death. Although infected individuals may not present any signs or symptoms of active disease, being infected with TB bacilli is a serious threat to the health status of the infected individual. Individuals who are infected have a 10% chance of developing active disease at some point in their life, a risk they would not have had without being infected. The risk of developing active disease is even greater for individuals who are immuno-compromised, due to any of a large number of factors. For example, individuals infected with HIV have been estimated as having an 8-10% risk *per year* of developing active disease (Ex. 4B).

In addition, since infected individuals commonly undergo treatment with anti-TB drugs to prevent the onset of active disease, they face the additional risk of serious side effects associated with the highly toxic drugs used to treat TB. Preventive treatment with isoniazid, one of the drugs commonly used to treat TB infection, has been shown in some cases to result in death from hepatitis or has damaged the infected person's liver to the extent that liver transplantation was performed (Ex. 6-10). Thus, the health hazards associated with TB infection clearly constitute material impairment of health.

Clinical illness, i.e., active disease, also clearly constitutes material impairment of health. Left untreated, 40 to 60 percent of active cases may lead to death (Exs. 7-50, 7-66, 7-80). Individuals with active disease may be infectious for various periods of time and often must be hospitalized. Active disease is marked by a chronic and progressive destruction of the tissues and organs infected with the bacteria. Active TB disease is usually found in the lungs (i.e., pulmonary tuberculosis). Long-term damage can result even when cases of TB are cured; a common result of TB is reduced lung function (impaired breathing) due to lung damage (Ex. 7-50, pp. 30-31). Inflammatory responses caused by the disease produce weakness, fever, chest pain, cough, and, when blood vessels are eroded, bloody sputum. Also, many individuals have drenching night sweats over the upper part of the body several times a week. The intensity of the disease varies, ranging from minimal symptoms of disease to massive involvement of many tissues, with extensive cavitation and debilitating constitutional and respiratory problems. Long-term damage can also result from extrapulmonary forms of active disease; such damage may include mental impairment from meningitis (infection of membranes surrounding the brain and spinal cord) and spinal deformity and leg weakness due to infection of the vertebrae (i.e., skeletal TB) (Ex. 7-50, p. 31). Active disease is treatable but it must be treated with potent drugs that have to be taken for long periods of time. The drugs currently used to treat active TB disease may be toxic to other parts of the body. Commonly reported side effects of anti-TB drugs include hepatitis, peripheral neuropathy, optic neuritis, ototoxicity and renal toxicity (Ex. 7-93). Active disease resulting from infection with MDR-TB is of even greater concern due to the inability to find adequate drug regimens. Although OSHA has not been able to precisely quantify the increase in incidence of MDR-TB, the number of cases of MDR-TB is clearly on the rise. In these cases, individuals may remain infectious for longer periods of time and may suffer more long-term damage from the chronic progression of the disease until adequate therapy can be identified.

In this standard, OSHA has presented quantitative estimates of the lifetime risk of TB infection, active disease and death from occupational exposure to *M. tuberculosis*. Qualitative evidence of occupational transmission is also included in OSHA's risk assessment.

In preparing its quantitative risk assessment, OSHA began by seeking out occupational data associated with TB infection incidence in order to calculate an estimate of risk for TB infection attributable to occupational exposure for all U.S. workers. Unfortunately, an overall national estimate of risk for TB infection attributable to occupational exposure is not available. CDC, which collects and publishes the number of active TB cases reported nationwide each year, does not publish occupational data associated with the incidence of TB infection and active TB on a nationwide basis. There has been some effort to include occupational information on the TB reporting forms, but only a limited number of states are currently using the new forms and capturing occupational information in a systematic way. In the absence of a national database, OSHA used two statewide studies, from North Carolina and Washington (Exs. 7-7, 7-263), and data from an individual hospital, Jackson Memorial Hospital (Ex. 7-108), on conversion rates of TB infection for workers in hospitals. The Washington State database also contained information on three additional occupational groups: long-term care, home health care and home care employees. OSHA used these data to model average TB infection rates and estimate the range of expected risks in the U.S. among workers with occupational exposure to TB.

The conversion rates in the selected studies were used to estimate the annual excess relative risk due to occupational exposure, which was expressed as the percent increase of infection above each study's control group. In order to estimate an overall range of occupational risk of TB infection, taking into account regional differences in TB prevalence in the U.S. and indirectly adjusting for factors such as socioeconomic status, which might influence the rate of TB observed in different parts of the country, OSHA: (1) Estimated background rates of infection for each state by assuming that the number of new infections is functionally related to the number of active cases reported by the state each year (i.e., the distribution of new infections is directly proportional to the distribution of active cases), and 2) applied estimates of the annual excess relative risk, derived from the occupational studies, to the state background rates to calculate estimates of excess risk due to occupational exposure by state. Thus, the excess occupational risk estimates are actually calculated from the three available studies, on a relative increase basis, and these relative increases are multiplied by background rates for each state to derive estimates of excess occupational risk by state. The state estimates are then used to derive a national estimate of annual occupational risk of TB infection. Given an annual rate of infection, the lifetime risk of infection was calculated assuming that workers are exposed for 45 years and that the worker's exposure profile and working conditions remain constant throughout his or her working lifetime. Lifetime infection rates are then used to calculate the lifetime risk of developing active disease based on the estimate that 10% of all infections result in active disease. Given a number of active cases of TB, the number of expected deaths can be calculated based on the estimated average TB case death rate (i.e., number of TB deaths per number of active TB cases averaged over 3 years as reported by CDC).

OSHA estimates that the risk of material impairment of health or functional capacity, that is, the average lifetime occupational risk of TB infection for hospital workers ranges from 30 to 386 infections per 1,000 workers who are occupationally exposed to TB. These are different national averages, each derived by calculating the risk in each state and weighting it by the state's population. The low end of this range is derived by using the Washington State data, and is likely to seriously underestimate the true risk to which workers are exposed. This is because the Washington data represent occupational exposures among employees in hospitals which are located in areas of the country with a low prevalence of active TB and which have implemented TB controls (e.g., early identification procedures, annual skin testing, and negative pressure in AFB isolation rooms). The high end of this range is derived by using the Jackson Memorial Hospital study, and represents occupational risk for workers in hospitals located in high TB prevalence areas, serving high risk patients, and with a high frequency of exposure to infectious TB.

OSHA also used information from the Washington State database to estimate national average estimates of lifetime risk for workers in long-term care (i.e., nursing homes), home health care, and home care. The national average lifetime risk of TB infection is estimated to be 448 per 1,000 for workers in long-term care facilities, 225 per 1,000 for workers in home health care (primarily nursing staff), and 69 per 1,000 for workers in home care. The higher likelihood of occupational exposure in long-term care facilities (early identification of suspect TB cases is often difficult among the elderly) and the presence of fewer engineering controls in these facilities may explain the high observed occupational risk in that work setting.

The national average lifetime risk of developing active disease ranges from approximately 3 to 39 cases per 1,000 exposed employees for workers in hospital settings. Similarly, the average lifetime risk of active disease is estimated to be approximately 45 per 1,000 for workers in long-term care, 26 per 1,000 in home health care, and 7 per 1,000 in home care. This range is based on the estimate that 10% of infections will progress to active disease over one's lifetime. This risk may be greater for immunocompromised individuals.

The national average lifetime risk of death from TB ranges from 0.2 to approximately 3 deaths per 1,000 exposed employees for workers in hospital settings. Similarly, the average lifetime risk of death from TB is estimated to be approximately 3.5 per 1,000 for workers in long-term care, 2 per 1,000 for workers in home health care, and 0.5 per 1,000 in home care. The lower range of the national lifetime risk of deaths, 0.2 per 1,000, is based on the Washington State hospital data where the prevalence of TB is low and infection control measures have been implemented. Thus, this lower range of risk underestimates the risk of death from TB for other employees who work in settings where infection control measures, such as those outlined in this proposed standard, have not been implemented. The risk assessment data show that where infection control measures were not in place, the estimated risk of death from TB was as high as 6 deaths per 1,000 exposed employees.

The quantitative risk estimates are based primarily upon data from hospitals and selected other work settings. However, it is frequent exposure to aerosolized *M. tuberculosis* which places workers at substantially increased risk of infection and not factors unique to the health care profession or any job category therein. Qualitative evidence, such as that from the epidemiological studies, case reports and outbreak investigations reported for various types of work settings, as discussed earlier in the Health Effects section, clearly demonstrates that employees exposed to aerosolized *M. tuberculosis* have become infected with TB and have gone on to develop active disease. These work settings share risk factors that place employees at risk of transmission. For example, these work settings serve client populations that are composed of a high prevalence of individuals who are infected with TB, are immunocompromised, are injecting drug users or are medically underserved and of poor general health status. Therefore, there is an increased likelihood that employees in these work settings will encounter individuals with active TB. In addition, high-hazard procedures, such as bronchoscopies, are performed in some of these work settings, which greatly increases the likelihood of generating aerosolized *M. tuberculosis*. Moreover, some of the work settings have environmental conditions such as overcrowding and poor ventilation, factors that facilitate the transmission of disease. Therefore, OSHA believes that the quantitative risk estimates based on hospital data and other selected health care settings can be extrapolated to other occupational settings where there is a similar increased likelihood of exposure to aerosolized *M. tuberculosis*.

Having specific data for non-health care workers and workplace conditions would add more precision to the quantitative risk assessment, but that level of detail is not possible with the currently available information. However, the Agency believes that such a level of detail is not necessary to make its findings of significant risk because the risk of infection is based upon occupational exposure to aerosolized *M. tuberculosis*. Nevertheless, OSHA seeks information on conversion rates and the incidence of active disease among employees in non-health care work settings in order to give more precision to its estimates of risk.

OSHA's risk estimates for TB infection are comparable to other risks which OSHA has concluded are significant, and are substantially higher than the example presented by the Supreme Court in the Benzene Decision. After considering the magnitude of the risk as shown by the quantitative and qualitative data, OSHA preliminarily concludes that the risk of material impairment of health from TB infection is significant.

OSHA also preliminarily concludes that the proposed standard for occupational exposure to TB will result in a substantial reduction in that significant risk. The risk of infection is most efficiently reduced by implementing TB exposure control programs for the early identification and isolation of individuals with suspected or confirmed infectious TB. Engineering controls to maintain negative pressure in isolation rooms or areas where infectious individuals are being isolated will reduce the airborne spread of aerosolized *M. tuberculosis* and subsequent exposure of individuals, substantially reducing the risk of infection. In addition, for those employees who must enter isolation rooms or provide services to individuals with infectious TB, respiratory protection will reduce exposure to aerosolized *M. tuberculosis* and thus reduce the risk of infection.

Several studies have shown that the implementation of infection control measures such as those outlined in this proposed standard have resulted in a reduction in the number of skin test conversions among employees with occupational exposure to TB. For example, results of a survey conducted by the Society of Healthcare Epidemiology of America (SHEA) of its member hospitals (Exs. 7-147 and 7-148) revealed that among hospitals that treated 6 or more patients with infectious TB per year there were 68% fewer tuberculin skin test conversions in hospitals that had AFB isolation rooms with one patient per room, negative pressure, exhaust air directed outside and six or more air changes per hour, compared to hospitals that did not have AFB isolation rooms with these same characteristics. Similarly, an 88% reduction in tuberculin skin test conversions was observed in an Atlanta hospital after the implementation of infection control measures such as an expanded respiratory isolation policy, improved diagnostic and testing procedures, the hiring of an infection control coordinator, expanded education of health care workers, increased frequency of tuberculin skin tests, implementation of negative pressure, and use of submicron masks for health care workers entering isolation rooms (Ex. 7-173). Improvements in infection control measures in a Florida hospital after an outbreak of MDR-TB reduced tuberculin skin test conversions from 28% to 18% to 0% over three years (Ex. 7-167). These improvements included improved early identification procedures, restriction of high-hazard procedures to AFB isolation rooms, increased skin testing, expansion of initial TB treatment regimens, and daily inspection of negative pressure in AFB isolation rooms. Thus, these investigations show that the implementation of infection control measures such as those included under OSHA's proposed standard for TB can result in substantial reductions in infections among exposed employees.

As discussed in further detail in the following section of the Preamble to this proposed standard, OSHA estimates that full implementation of the proposed standard for TB will result in avoiding approximately 21,400 to 25,800 work-related infections per year, 1,500 to 1,700 active cases of TB resulting from these infections and 115 to 136 deaths resulting from these active cases. In addition, because the proposed standard encourages the identification and isolation of active TB cases in the client populations served by workers in the affected industries, there will also be non-occupational TB infections that will be averted. OSHA estimates that implementation of the proposed standard will result in avoiding approximately

3,000 to 7,000 non-occupational TB infections, 300 to 700 active cases of TB resulting from these infections, and 23 to 54 deaths resulting from these active cases. OSHA preliminarily concludes that the proposed standard for TB will significantly reduce the risk of infection, active disease and death from exposure to TB and that the Agency is thus carrying out the Congressional intent and is not attempting to reduce insignificant risks.

Although the current OSHA enforcement program, which is based on the General Duty Clause of the Act, Section 5(a)(1), and the application of some general industry standards, such as 29 CFR 1910.134, Respiratory Protection, has reduced the risks of occupational exposure to tuberculosis to some extent, significant risks remain and it is the Agency's opinion that an occupational health standard promulgated under section 6(b) of the Act will much more effectively reduce these risks for the following reasons. First, because of the standard's specificity, employers and employees are given more guidance in reducing exposure to tuberculosis. Second, it is well known that a standard is more protective of employee health than an enforcement program based upon the general duty clause and general standards. Unlike the proposed standard, the general duty clause specifies no abatement methods and the general industry standards do not set forth abatement methods specifically addressing occupational exposure to TB. Third, the general duty clause imposes heavy litigation burdens on OSHA because the Agency must prove that a hazard exists at a particular workplace and that it is recognized by the industry or the cited employer. Since the proposed standard specifies both the conditions that trigger the application of the standard and the employer's abatement obligations, thereby establishing the existence of the hazard, no independent proof that the hazard exists in the particular workplace need be presented. The reduction in litigation burdens will mean that the Labor Department, as well as the employer, will save time and money in the investigation and litigation of occupational TB cases. Finally, the promulgation of this proposed standard will result in increased protection for employees in state-plan states which, although not required to adopt general duty clauses, must adopt standards at least as effective as Federal OSHA standard.

In summary, the institution of the enforcement guidelines has been fruitful, but it has not eliminated significant risks among occupationally exposed employees. Therefore, OSHA preliminarily concludes that a standard specifically addressing the risks of tuberculosis is necessary to further substantially reduce significant risk. OSHA's preliminary economic analysis and regulatory flexibility analysis indicate that the proposed standard is both technologically and economically feasible. OSHA's analysis of the technological and economic feasibility is discussed in the following section of the preamble.

SUMMARY OF THE PRELIMINARY ECONOMIC ANALYSIS AND REGULATORY FLEXIBILITY ANALYSIS

OSHA is required by the Occupational Safety and Health Act of 1970 and several court cases pertaining to that Act to ensure that its rules are technologically and economically feasible for firms in the affected industries. Executive Order (EO)

12866 and the Regulatory Flexibility Act (as amended) also require Federal agencies to estimate the costs, assess the benefits, and analyze the impacts on the regulated community of the regulations they propose. The EO additionally requires agencies to explain the need for the rule and examine regulatory and non-regulatory alternatives that might achieve the objectives of the rule. The Regulatory Flexibility Act requires agencies to determine whether the proposed rule will have a significant economic impact on a substantial number of small entities, including small businesses and small government entities and jurisdictions. For proposed rules with such impacts, the agency must prepare an Initial Regulatory Flexibility Analysis that identifies those impacts and evaluates alternatives that will minimize such impacts on small entities. OSHA finds that the proposed rule is "significant" under Executive Order 12866 and "major" under Section 804(2) of the Small Business Regulatory Enforcement Fairness Act of 1996. Accordingly, the Occupational Safety and Health Administration (OSHA) has prepared this Preliminary Economic and Regulatory Flexibility Analysis (PERFA) to support the Agency's proposed standard for occupational exposure to tuberculosis (TB). The following is an executive summary of that analysis. The entire test of the PERFA can be found in the rulemaking docket as Exhibit 13. The complete PERFA is composed of various chapters that describe in detail the information summarized in the following section.

STATEMENT OF NEED

TB is a communicable, potentially lethal disease caused by the inhalation of droplet nuclei containing the bacillus *Mycobacterium tuberculosis* (*M. tuberculosis*). Persons exposed to these bacteria can respond in different ways: by overcoming the challenge without developing TB, by becoming infected with TB, or by developing active TB disease. Those who become infected harbor the infection for life, and have a 10 percent chance of having their infection progress to active disease at some point in their life. Those with active disease have the signs and symptoms of TB (e.g., prolonged, productive cough; fatigue; night sweats; weight loss) and have about an 8 percent risk of dying from their disease.

TB has been a worldwide health problem for centuries, causing millions of deaths worldwide. In the United States, however, there has been a decline in the number of active TB cases over the last four decades. Between 1953 and 1994, the number of active cases declined from 83,304 to 24,361, an annual rate of decline of 3.6 percent over the period as a whole (Figure VII-1). The 1988-1992 period, however, saw the first substantial increase in the number of active cases since 1953. A number of outbreaks of this disease have occurred among workers in health care settings, as well as other work settings, in recent years. To add to the seriousness of the problem, some of these outbreaks have involved the transmission of multi-drug resistant strains of *M. tuberculosis*, which are often fatal. Very recently, i.e., after 1992, this trend has reversed, and the number of such cases appears once again to have begun to decline. Nevertheless, TB remains a major health problem, with 22,813 active cases reported in 1995. Because active TB is endemic in many U.S. populations—including groups in both urban and rural areas—workers who come into contact with diseased individuals are at risk of contracting the disease themselves.

6 Hepatitis C in Health Care Workers

Jon Rosenberg

CONTENTS

BACKGROUND

In 1975, a type of hepatitis was described that occurred in recipients of blood transfusions and was clearly not type A or B; this came to be known as "parenterally transmitted non-A, non-B hepatitis." The diagnosis was made after exclusion of identifiable causes of liver disease. Although it took 15 years to identify the hepatitis C virus (HCV), non-A, non-B hepatitis (NANBH) was characterized clinically as having a relatively long incubation period (approximately 60 days); a mild, usually subclinical presentation; and a high rate of progression to cirrhosis. It was characterized epidemiologically as being responsible for approximately 40% of all infectious hepatitis in the United States and for most of the hepatitis that followed blood transfusions in approximately 15% of recipients. The use of surrogate markers for infection (hepatitis B antibodies and abnormal liver function), and finally the development of HCV antibody tests, resulted in a dramatic lowering of the incidence of posttransfusion hepatitis.

THE HEPATITIS C VIRUS

The HCV is a small (30–60-nm) single-stranded RNA virus of the Flaviviridae family. It is sensitive to organic solvents, indicating a lipid envelope. It appears to be genetically diverse, with a definite geographic distribution to identified subtypes. The pathogenesis of hepatic injury is believed to be immunologic. The immunologic response does not appear to be protective against either chronic infection or future infection with additional strains of the virus, possibly as a result of its genetic diversity.

HEPATITIS C VIRUS TESTS

ANTIBODY TESTS

Tests that detect antibody against the virus include the enzyme immunoassays (EIAs), and the recombinant immunoblot assays (RIBAs). The RIBAs contain the same HCV antigens as EIA, and so are supplementary to the EIA, which is used as an initial screening test, rather than a confirmatory test. The EIA has moved from a first-generation test of limited sensitivity and poor specificity, to the second-generation test (EIA-2), which has a 92–95% sensitivity, using tests for HCV RNA as a standard. Its specificity is uncertain. About 25–60% of blood donors with no risk factors for hepatitis C who are positive by the EIA-2 test are also positive by the polymerase chain reaction (PCR) test for HCV RNA. Of low-risk donors who are both EIA-2 and RIBA-positive, 70–75% are positive for HCV RNA. Positive predictive values are much higher in patients with hepatitis C risk factors, elevated liver enzyme serum alanine aminotransferase (ALT) levels, or clinical liver disease. A detailed discussion of the diagnosis of hepatitis C is available in Reference 1.

POLYMERASE CHAIN REACTION

The HCV nucleic acid sequence can be amplified by PCR, enabling the detection of viremia (virus in blood) in patients with HCV infection. This has proved a powerful research tool, but is unfortunately available only to a limited extent for clinical use. Viremia can be detected as early as 3 days after exposure. The persistence of viremia is predictive of progression to chronic hepatitis, even with normal ALT values, while the disappearance of viremia correlates with the resolution of disease. However, even PCR results must be interpreted with caution. A quality control study of 31 hepatitis C research laboratories in Europe, the United States, and Japan, found that only 5 (16%) performed flawlessly with the entire panel of samples, and even these 5 reported a 100-fold difference in sensitivity. False positive results due to contamination were the greatest problem: 11% of the negative samples were reported positive. Clinicians should be aware of the proficiency record of laboratories performing HCV RNA testing to ensure test accuracy for their patients.

EPIDEMIOLOGY OF HEPATITIS C

HCV is the agent that most often causes NANBH in the United States, accounting for 20–40% of acute viral hepatitis (about equal to that of hepatitis A and B each). Worldwide, anti-HCV prevalence ranges from 0.3–1.5%; about 1% of volunteer blood donors are positive. Nearly 4 million Americans are infected with hepatitis C. Currently, approximately 30,000 acute new infections are estimated to occur each year, about 25–30% of which are diagnosed. Hepatitis C accounts for 20% of all cases of acute hepatitis. Currently, hepatitis C is responsible for an estimated 8,000–10,000 deaths annually, and without effective intervention that number is postulated to triple in the next 10–20 years. Hepatitis C is now the leading reason for liver transplantation in the United States.

Over half the new HCV infections each year in the United States are thought to be the result of injection drug use; the remainder are the result of transfusion prior to 1990, high-risk sexual activity, hemodialysis, and occupational exposure to blood. The route of transmission is unknown in less than 10% of cases.

CLINICAL FEATURES AND NATURAL HISTORY OF HEPATITIS C

ACUTE HEPATITIS C

After initial exposure, HCV RNA can be detected in blood in 1–3 weeks. Within an average of 50 days (range: 15–150 days), virtually all patients develop liver cell injury, as shown by elevation of ALT. Acute illness is usually mild and onset is insidious. The majority of patients have no symptoms or jaundice; about 25% develop malaise, weakness, or anorexia. Fulminant liver failure is rare. Antibodies to HCV (anti-HCV) almost invariably become detectable during the course of illness, appearing in approximately 90% of patients 3 months after onset of infection. In about 15% of cases there is complete recovery with disappearance of HCV RNA from blood and return of liver enzymes to normal.

CHRONIC HEPATITIS C

About 85% of HCV-infected individuals fail to clear the virus by 6 months and develop chronic hepatitis. This capacity to produce chronic hepatitis is one of the most striking features of HCV infection. The majority of patients with chronic infection have abnormalities in ALT levels that can fluctuate widely. About one third of patients have persistently normal serum ALT levels. Antibodies to HCV or circulating viral RNA can be demonstrated in virtually all patients.

Chronic hepatitis C is typically an insidious process, progressing, if at all, at a slow rate without symptoms or physical signs in the majority of patients during the first 20 years. A small proportion of patients with chronic hepatitis C—perhaps less than 20%—develop nonspecific symptoms, including mild intermittent fatigue and malaise. Symptoms first appear in many patients with chronic hepatitis C at the time of development of advanced liver disease.

Within 20 years about 20% of those infected develop cirrhosis, which is fibrosis (scarring) of the liver tissue. Progression to cirrhosis cannot be reliably predicted by clinical or biochemical features; it occurs more rapidly in those who drink alcohol. After 20 years of chronic hepatitis C between 1 and 5%, usually those with cirrhosis, develop liver cancer, which is almost invariably fatal.

TREATMENT OF HEPATITIS C

Chronic Hepatitis

The principal treatment of hepatitis C is with interferon alfa, although other forms of interferon appear to be equally efficacious. The efficacy of interferon therapy currently is defined biochemically as normalization of serum ALT and virologically as loss of serum HCV RNA. Based on these markers, randomized clinical trials have demonstrated that treatment with interferon benefits some patients with chronic hepatitis C. Treatment with interferon at a dosage of 3 million units administered subcutaneously three times weekly for 6 months has produced a biochemical and virological responses in 30–50% of patients at the end of treatment, but about half of these relapse after treatment is stopped.

Increasing the duration of treatment to 12 months somewhat improves the response after treatment. The benefit of treatment of longer duration is still being evaluated. It should be recognized that although interferon treatment may be associated with favorable effects on biochemical and virological markers, its effects on important clinical outcomes such as quality of life and disease progression remain undetermined. Because most patients do not experience sustained response, attempts have been made to identify individuals who are more likely to respond to therapy. The important factors associated with a favorable response to treatment include HCV genotype 2 or 3, low serum HCV RNA level, and absence of cirrhosis.

Interferon causes both bothersome side effects and serious toxicity. A reduction in interferon dosage is required in 10–40% of patients because of side effects, and treatment must be discontinued in 5–10%. Rare deaths have occurred, principally in patients with cirrhosis. A percutaneous liver biopsy is usually obtained before initiating therapy with interferon. Frequent monitoring during therapy is required.

The additional drug of most promise, at present, is ribavirin, an oral antiviral agent that, when used alone, reduces serum ALT levels in approximately 50% of patients. However, ribavirin by itself does not lower serum HCV RNA levels, and relapses occur in virtually all patients when therapy is stopped. Of greater promise are reports that the combination of interferon and ribavirin leads to higher sustained virological response rates (40–50%) than interferon alone in 6-month clinical trials. Large-scale trials of the combination are now under way.

All patients with chronic hepatitis C are potential candidates for specific therapy. However, given the current status of therapies for hepatitis C, treatment is clearly recommended only in a selected group of patients. In others, treatment decisions are less clear and should be made on an individual basis or in the context of clinical trials.

ACUTE HEPATITIS

Preliminary data suggest a benefit from interferon treatment with higher clearance of HCV RNA in patients with acute hepatitis C. According to the National Institutes of Health (NIH) consensus statement, in light of these findings, interferon treatment of patients with acute hepatitis C could be recommended. This is particularly relevant for health care workers, in whom the diagnosis of acute hepatitis C could be established by the monitoring of acute symptoms (which occur in 25–35% of those infected) following exposure or, more effectively, by the weekly measurement of HCV RNA, which will be detected in 1–3 weeks in essentially all who become infected. Based on its putative mechanisms of action, it is not thought that interferon would likely be of benefit if administered as post-exposure prophylaxis, prior to the development of infection.

TRANSMISSION OF HCV

HCV circulates in low titers in the blood of infected persons and is detected inconsistently in other body fluids. The most efficient mode of HCV transmission is direct percutaneous blood exposure, such as through the transfusion of blood or blood products or the transplantation of organs from infected donors and through the sharing of contaminated needles among injection-drug users. The risk for HCV transmission after a needlestick exposure to blood from a source positive for anti-HCV is 3–10%, compared to 6–30% for HBV and 0.3% for HIV. The risk for acquiring HCV infection from a blood transfusion is now approximately 0.5% per patient (0.03% per unit). Other bloodborne viruses, such as the hepatitis B virus (HBV), are transmitted not only by overt percutaneous exposures, but also by mucous membrane and inapparent parenteral exposures. Although these types of exposures are prevalent among health care workers, the risk factors for HCV transmission from them are unknown.

A small number of epidemiological studies demonstrate that perinatal, sexual, and household transmission occurs, but our understanding of the risks of transmission in these settings has been limited by inadequate studies, the need for more sensitive tests to detect infection, and the inability to quantitate infectivity.

RISK OF OCCUPATIONAL HCV TRANSMISSION

The risk of HCV transmission is related to: (1) the likelihood virus is present in source (i.e., the prevalence of HCV infection in patients); (2) the level of virus in exposure medium (i.e., the level of viremia in the patient's blood); (3) the route, nature, and degree of exposure; and (4) the type and size of the inoculum.

HCV antibodies have been found in a range 2–34% of patients in the United States and United Kingdom. Kelen et al.[7] found that of all adult emergency room (ER) patients (2523) at Johns Hopkins Hospital, 18% had antibodies against HCV, compared to 5% for HBV surface antigen and 6% for human immunodeficiency virus (HIV) antibodies. Evans et al.[8] at San Francisco General Hospital found that of 88 source patients tested, 34% had antibodies against HCV, compared to 2.3% for HBV surface antigen and 22.45% for HIV antibodies. Zuckerman and co-workers[11] found in a London hospital that 24 (13.9%) of 173 source patients tested positive for anti-HCV antibodies. In all these studies, from 25–67% of the source patients, the HCV infection was unknown prior to testing.

Evidence for occupational transmission of HCV comes from case reports, sero-prevalence surveys, and seroconversion studies. Case reports have demonstrated the existence of occupational transmission of HCV, but do not estimate the risk of transmission. Seroprevalence surveys can estimate the risk of transmission based on comparison of occupational groups to control groups, but the estimation of risk is critically dependent on the selection of comparison groups and the participation rates in both groups. Seroconversion studies can provide the best estimate of risk, but require a large study population, long time, and high degree of acertainment; and are critically dependent on the selection of the denominator (i.e., what exposure is, what a needlestick is, and what type the needlestick is). Since current antibody tests detect 90% of infections, their use in such studies will detect the majority, but not all, of occupational infections.

SEROPREVALENCE STUDIES

There have been 14 seroprevalence studies in hospital health care workers reported to date. Of five U.S. studies from 1991 to 1995, prevalences of 1.4–2.0% were reported, compared to a volunteer blood donor prevalence of 0.5%. Dialysis workers and dentists are known to be at higher risk of HBV infection than are other health care workers. Two studies of dialysis health care workers found prevalence of 2.5% in Italy and 4.1% in Belgium. Two studies were done of dentists in the United States, with prevalence of 0.7–1% found in dentists and of 2–9.3% found in oral surgeons. While volunteer blood donors are usually used as the comparison group in such studies, it is likely that the prevalence in donors, usually less than 1%, is an under-estimate of the nonoccupational prevalence in general health care worker populations (due to the presence of risk factors for HCV infection that would preclude some workers from donating blood). The prevalence in the general U.S. population is currently 1.8%.

NEEDLESTICK SEROCONVERSION STUDIES

There are nine studies of follow-up of health care workers after percutaneous exposure to blood from HCV-positive patients reported to date. Five studies have detected seroconversion after needlesticks. One U.S. study reported a seroconversion rate of 3 of 50 (6%) exposures. In three Japanese studies, 13/276 (3–9%) workers seroconverted; in one using PCR, the rate was 10%. An Italian study involving 16 hospitals reported that 4 (1.2%) of 331 hollow-bore HCV-positive needlesticks resulted in infection. Four studies have reported no seroconversions after exposure, including two in Italy, one in the United Kingdom, and one in Spain, including a total of 146 health care worker exposures.

The following comparisons can be made among HBV, HCV, and HIV:

| | Percentage Testing Positive | | |
Population	HBV	HCV	HIV
Blood donors	0.03 (HBsAg)[a]	0.5	0.005–0.02
Patients (inner city)	2–6 (HBsAg)[a]	15–35	2–20
Health care workers	10–30 (all markers)	1–2	<1
Health care workers following needlestick	10–60	0–10	0.5

[a] HBsAg, hepatitis B surface antigen.

SUMMARY OF OCCUPATIONAL HEPATITIS C VIRUS

The occupational risk of HCV to health care workers is low relative to HBV, and higher than HIV. Rates should only be compared to each other, and not described as "low" or "high".

The risk of transmission of HCV from a contaminated needlestick is between that for HBV and for HIV.

- Risk is probably related to level of viremia in blood of source patient, size of inoculum, and depth of needlestick.

Other risk factors for HCV transmission in health care settings are not well defined, but are probably similar to those for HIV and HBV.

Risk for nonparenteral exposure is uncertain.

- A single case of seroconversion has been reported from conjunctival blood splash.

EXPOSURE FOLLOW-UP ISSUES

According to the CDC:[5]

HCV cannot be prevented by immunization.

- Antibodies produced by infection are not neutralizing.
- Infection with additional strains occurs.

HCV transmission cannot be reduced by postexposure prophylaxis.

- Immunoglobulin (Ig) is now screened for anti-HCV.
- Screened Ig is not protective in chimpanzees administered Ig 1 h after HCV exposure.

Data on risk of transmission are limited.

There are limitations of available serological testing for detecting infection and determining infectivity.

Risks of transmission by sexual, household, and perinatal exposures are poorly defined.

There are no recommendations for changes in sexual practice, pregnancy, or breastfeeding.

- Infected persons are informed of possible risk

There is limited benefit of therapy for chronic disease.

The cost of follow-up testing or treatment is a consideration.

Medical–legal implications include the effects on health care workers' careers and workers' compensation.

There is a risk of occupational transmission.

- What if any is the risk of nonpercutaneous exposure?

Testing limitations include

- The prolonged interval between exposure and seroconversion varies from an average of 8–10 weeks to occasionally longer than 6 months.
- As to specificity, false positives lead to supplemental tests.
- A sensitivity of 90–95% can be determined by EIAs.
- PCR is not standardized, and has a high cost ($200) and uncertain reliability.

RECOMMENDATIONS

According to the CDC:[5]

1. No postexposure prophylaxis is available for hepatitis C; Ig is not recommended.
2. Institutions should provide to health care workers accurate and up-to-date information on the risk and prevention of all bloodborne pathogens, including hepatitis C.
3. Institutions should consider implementing policies and procedures for follow-up of health care workers after percutaneous or per mucosal exposure to anti-HCV positive blood. Such policies might include baseline testing of the source for anti-HCV and baseline and 6-month follow-up testing of the person exposed for anti-HCV and ALT activity. All anti-HCV results reported as repeatedly reactive by EIA should be confirmed by supplemental anti-HCV testing.
4. There are currently no recommendations concerning restriction of health care workers with hepatitis C. The risk of transmission from an infected worker to a patient appears to be very low. Furthermore, there are no serological assays that can determine infectivity and no data to determine the threshold concentration of virus required for transmission. As recommended for all health care workers, those who are anti-HCV positive should follow strict aseptic technique and standard (universal) precautions, including appropriate use of hand washing, protective barriers, and care in the use and disposal of needles and other sharp instruments.

CONCLUSIONS

The "low" risk of transmission does not adequately reflect the concern over exposure on the part of health care workers, particularly over the lack of immunization protection (which exists for hepatitis B) or postexposure prophylaxis (which exists for HIV). Health care workers also are concerned about possible loss of jobs or careers if the HCV-positive status becomes known. These concerns have led health care workers to not report exposure incidents to known HCV-positive patients, or to decline opportunities to be tested themselves.

Therefore, greater attention should be paid to recognition of and postexposure follow-up for exposure incidents, including

- Source testing
- Worker testing
- Counseling
- Medical evaluation for infected workers
- Medical treatment for infected workers
- Addressing employment security for infected workers

Needlestick prevention is likely to be the only effective means of preventing occupational HCV infection for the immediate future.

REFERENCES

GENERAL

1. National Institutes of Health Consensus Development Panel. Management of hepatitis C. National Institutes of Health, March 27, 1997.
2. Mast EE, Alter MJ. Hepatitis C. *Semin Pediatr Infect Dis.* 1997;8:17–22.
3. Alter MJ. Epidemiology of hepatitis C in the West. *Semin Liver Dis* 1995;15:5–14.
4. Centers for Disease Control. Public Health Service inter-agency guidelines for screening donors of blood, plasma, organs, tissues, and semen for evidence of hepatitis B and hepatitis C. *MMWR* 1991;40(RR-4):13–14.

OCCUPATIONAL

5. Centers for Disease Control and Prevention. What is the risk of acquiring hepatitis C for healthcare workers and what are the recommendations for prophylaxis and follow-up after occupational exposure to hepatitis C?. Hepatitis Surveillance Report No. 56. Atlanta: Centers for Disease Control and Prevention, 1995, pp. 3–6.
6. Polish LB, Tong MJ, Co RL, et al. Risk factors for hepatitis C virus infection among health care personnel in a community hospital. *Am J Infect Control* 1993;21: 196–200.

PREVALENCE IN PATIENTS

7. Kelen GD, Green GB, Purcell RH, et al. Hepatitis B and hepatitis C in emergency department patients. *N Engl J Med* 1992;326:1399–1404.
8. Evans SE, Fahrner R, Gerberding JL. HIV, HBV, and HCV prevalence among source patients reported to the San Francisco General Hospital hotline. Presented at the 34th Annual Meeting of the Infectious Disease Society of America; September 18–20, 1996; New Orleans, LA. Abstract 24.
9. Louie M, Low DE, Feinman SV, et al. Prevalence of bloodborne infective agents among people admitted to a Canadian hospital. *Can Med Assoc J* 1992;146:1331–1334.

10. Thomas DL, Cannon RO, Shapiro CN, et al. Hepatitis C, hepatitis B, and human immunodeficiency virus infections among non-intravenous drug-using patients attending clinics for sexually transmitted diseases. *J Infect Dis* 1994;169:990–995.

PREVALENCE IN HEALTH CARE WORKERS

11. Zuckerman J, Clewley G, Griffiths P, Cockcroft A. Prevalence of hepatitis C antibodies in clinical health-care workers. *Lancet* 1994;343:1618–1620.
12. Petrosillo N, Puro V, Ippolito G, and Italian Study Group on Blood-borne Occupational Risk in Dialysis. Prevalence of hepatitis C antibodies in health-care workers. *Lancet* 1994;344:339–340.
13. Neal KR, Dornan J, Irving WL. Prevalence of hepatitis C antibodies among healthcare workers of two teaching hospitals. Who is at risk? *Br Med J* 1997;314:179–180.
14. Zuckerman J, Clewley G, Griffiths P, et al. Prevalence of hepatitis C antibodies in clinical health-care workers. *Lancet* 1994;343:1618–1620.
15. Struve J, Aronsson B, Frenning B, et al. Prevalence of antibodies against hepatitis C virus infection among health care workers in Stockholm. *Scand J Gastroenterol* 1994;29:360–362.
16. Libanore M, Bicocchi R, Ghinelli, et al. Prevalence of antibodies to hepatitis C virus in Italian health care workers. *Infection* 1992;20:50.
17. Jadoul M, El Akrout M, Cornu C, et al. Prevalence of hepatitis C antibodies in health-care workers. *Lancet* 1994;344:339.
18. Petrarulo F, Maggi P, Sacchetti A, et al. HCV infection occupational hazard at dialysis units and virus spread, among relatives of dialyzed patients. *Nephron* 1992;61:302–303.
19. Thomas DL, Gruninger SE, Siew C. Occupational risk of hepatitis C infections among general dentists and oral surgeons in North America. *Am J Med* 1996;100:41–45.
20. Klein RS, Freeman K, Taylor PE, et al. Occupational risk for hepatitis C virus infection among New York City dentists. *Lancet* 1991;338:1539-1542.

SEROCONVERSION STUDIES

21. Lanphear BP, Linnemann CC, Cannon CG, et al. Hepatitis C virus infection in health care workers: risk of exposure and infection. *Infect Control Hosp Epidemiol* 1994;15:745–750.
22. Puro V, Petrosillo N, Ippolito, et al. Occupational hepatitis C virus infection in Italian health care workers. *Am J Public Health* 1995;85:1272–1275.
23. Puro V, Petrosillo N, Ippolito. Risk of hepatitis C seroconversion after occupational exposure in health care workers. *Am J Infect Control* 1995;23:273–277.
24. Mitsui T, Iwano K, Masuko K, et al. Hepatitis C virus infection in medical personnel after needlestick accident. *Hepatology* 1992;16:1109–1114.
25. Kiyosawa K, Sodeyama T, Tanaka E, et al. Hepatitis C in hospital employees with needlestick injuries. *Ann Int Med* 1991;115:367–369.
26. Hernandez ME, Bruguera M, Puyuelo T, et al. Risk of needle-stick injuries in the transmission of hepatitis C virus in hospital personnel. *J Hepatol* 1992;16:56–58.
27. Sartori M, La Terra G, Aglietta M, et al. Transmission of hepatitis C via blood splash into conjunctiva. *Scand J Infect Dis* 1993;25:270–271.

DIAGNOSIS

28. Zaaijer HL, Cuypers HTM, Reesink HW et al. Reliability of polymerase chain reaction for detection of hepatitis C virus. *Lancet* 1993;341:722–724.

NONOCCUPATIONAL TRANSMISSION

29. Alter MJ, Coleman PJ, Alexander WJ, et al. Importance of heterosexual activity in the transmission of hepatitis B and non-A, non-B hepatitis. *JAMA* 1989;262:1201–1205.
30. Ohto H, Terazawa S, Sasaki N, et al. Transmission of hepatitis C virus from mothers to infants. *N Engl J Med* 1994;330:744–750.
31. Lin HH, Kao JH, Hsu HY, et al. Possible role of high-titer maternal viremia in perinatal transmission of hepatitis C virus. *J Infect Dis* 1994;169:638–641.

TREATMENT

32. Hoofnagle, JH, Di Bisceglie AM. The treatment of chronic viral hepatitis. *New Engl J Med* 1997;336:347–355.
33. Fried MW, Hoofnagle, JH. Therapy of hepatitis C. *Semin Liver Dis* 1995;15:82–91.

IMMUNOGLOBULIN POSTEXPOSURE PROPHYLAXIS

34. Krawczynski K, Alter MJ, Tankersley DL, et al. Effect of immune globulin on the prevention of experimental hepatitis C virus infection. *J Infect Dis* 1996.

INTERFERON FOR ACUTE HEPATITIS C

35. Schiff ER. Hepatitis C among health care providers: risk factors and possible prophylaxis. *Hepatology* 1992;16:1300–1301.
36. Noguchi S, Sata M, Suzuki H, et al. Early therapy with interferon for acute hepatitis C acquired through a needlestick. *Clin Infect Dis* 1997;24:992–994.
37. Omata M, Yokosuka O, Takano S, et al. Resolution of acute hepatitis C after therapy with natural beta interferon. *Lancet* 1991;338:914–915.
38. Lampertico P, Rumi M, Romeo R, et al. A multicenter randomized controlled trial of recombinant interferon-α_{2b} in patients with acute transfusion-associated hepatitis C. *Hepatology* 1994;19:19–22.
39. Viladomiu L, Genesca J, Esteban JI, et al. Interferon- in acute posttransfusion hepatitis C: a randomized, controlled trial. *Hepatology* 1992;15:767–769.

7 Epidemiology of Latex Allergy

George W. Weinert

CONTENTS

INTRODUCTION

Latex allergy is caused by a number of natural products known as antigens. These are found in the sap of the rubber tree. Some chemicals used to make rubber products can cause physiological reactions such as skin irritation, which may be reported as an allergy. The composition and concentration of the antigen and other irritating chemicals in rubber products are not uniform. Rubber, or latex sap, is produced in many areas of the world, though primarily in Malaysia. Adequate purification processes exist to remove the undesirable antigenic proteins, but often these processes are not used. A basic knowledge of these materials as well as consideration of the route of entry is important to the understanding of the epidemiology of latex allergy.

BACKGROUND

Why has this disease occurred suddenly in our society? The Occupational Safety and Health Administration (OSHA) began enforcing universal precautions in the late 1980s on the advice of the Centers for Disease Control and Prevention (CDC) in response to health care worker concern related to the acquired immunodeficiency syndrome (AIDS) epidemic. This OSHA enforcement activity required institutions to use personal protective methods to safeguard the workforce. This preliminary regulatory activity often resulted in citations by OSHA under the general duty clause. Eventually stronger regulatory activity occurred with the publication of the OSHA bloodborne pathogen standard (Code of Federal Regulation [CFR] Part 29, 1910.1030) published in the *Federal Register,* December 6, 1991. This required

0-8493-3382-2/99/$0.00+$.50

increased use of personal protective equipment, especially gloves. This prompted glove manufacturers to increase production. Glove productivity was doubled within 1 year, and the production of gloves continues to grow into 1998.

In March of 1991, the Food and Drug Administration (FDA) issued an alert to all hospitals concerning latex allergies. The FDA indicated it had received 16 complaints in 1989 and 1600 complaints in 1990 connected with latex problems. Several patient deaths directly related to latex were also reported by the FDA.

Increased consumer demand prompted an increase in the production of latex products. This consumer demand resulted in lower quality latex products emerging on the market (i.e., gloves and other products with a higher protein antigen content), and these lower quality products caused several responses. Literature references dating back 30 years or more identify that systemic latex allergy was known to the rubber industry.

There are three types of responses associated with latex glove use in health care. The first type of response is a simple chemical irritation of the skin; the second type of response is a cell-mediated allergic response that produces a more severe skin reaction; and the third type, a systemic allergic response that it is far more serious.

The simple chemical irritation skin response is caused by chemicals known as accelerators that are added to the latex sap to aid in the film-forming process. The cell-mediated allergic response is caused by the proteins found in latex sap. This is a true allergic response and is confined to the area of contact with the rubber product. This is a different response from that caused by the accelerators. The proteins found in latex sap also cause the systemic allergic response. This response may be manifested by asthma, allergic rhinitis, or anaphylaxis. This is an immunoglobulin E (IgE)-mediated response, and involves the entire body. The proteins that cause the cell-mediated allergic response and the systemic response are contaminates in the latex sap taken from the *Hevea brasiliensis* tree. Many of these same proteins have been identified in other rubber plants.

Glove removal and washing hands quickly reverses the simple chemical irritation. The cell-mediated allergic response may continue for several days and peak at about 48 h after exposure. However, the tissue will eventually return to normal with no further exposure. In both the simple chemical irritation and the cell-mediated allergic response, the skin reaction is confined to the area of contact with the gloves. There is currently no evidence that the cell-mediated allergic response, which is a true allergic response, will lead to a systemic allergic response. The systemic allergic response is an IgE-mediated response. These two mechanisms of response are both different than and independent of each other.

The last type of response, systemic allergic response is by far the most serious reaction. The symptoms of systemic allergic response are asthma, allergic rhinitis, and anaphylaxis (which may result in death). Dermal exposure can result in the chemical irritation or the cell-mediated allergic response, which is far less serious. Inhalation or invasive procedures with gloves or other latex-containing medical devices can result in sensitization and produce systemic allergic response, although the literature references are not always clear concerning this sensitization. The following discussion will be exclusively devoted to the epidemiology of systemic allergic response.

TABLE 1
Sensitization to Natural Rubber Latex

Response Type	Cause	Affected Group	Response Time	Permanency	Route of Exposure	Seriousness
Chemical irritation	Chemical additives	Care providers	Immediate	Transitory	Skin content	Not especially
Cell-mediated allergic response	Latex proteins	Care providers	48-h peak	Transitory	Skin content	Not life threatening
Systemic allergic response	Latex proteins	Care providers and patients	Delayed or immediate	Years	Inhalation and invasive contact	Can be life threatening

There are two routes of entry for the latex antigens, inhalation and invasive contact. The high-risk groups include individuals who have had multiple surgical procedures, rubber workers in tire manufacturing plants, and health care workers. Contact with the mucosa is the primary route of entry for patients. If we prevent high antigen content medical devices, such as gloves, from contact with the mucosa, exposure is eliminated. The FDA identified Foley catheters as a contact source of the latex antigen for patient sensitization. Barium enema retention rings were also identified and were the cause of all the deaths reported by the FDA.

Health care workers are sensitized by the inhalation of latex-containing dust most often related to glove use. If we carefully eliminate inhalation as a route of entry, then the potential for health care workers' sensitization is eliminated. The elimination of exposure by inhalation seems straightforward. However, the concentration for sensitization is in the nanograms of antigen per cubic meter of air range: 1 ng is 0.000000001 g, 0.0000001 mg, or 0.001 µg dispersed in a cubic meter of air. Some unusual techniques are necessary to control exposure. These techniques can be used effectively in the elimination of new cases of sensitization.

The author's personal experience while Director of Environmental, Safety and Health at Brigham and Women's Hospital in Boston, MA demonstrated that new cases of health care worker sensitization could be avoided by elimination of the inhalation route of entry.

The response types are further identified in Table 1. The National Institute for Occupational Safety and Health (NIOSH) published "Preventing Allergic Reactions to Natural Rubber Latex in the Work Place" in the June 1997, NIOSH publication 97-135. This document is available through the Internet, and it provides extensive information to help understand latex allergy. The publication also contains several pertinent references.

Health care worker sensitization is caused by the inhalation of latex protein particles dispersed in the air. These protein particles are thought to be attached to the powder of powered gloves. Lower powdered gloves, in general, present less of a problem than heavily powdered gloves. However, these protein particles may also be dispersed in the air from nonpowdered gloves. The key to elimination of new sensitization is to eliminate the protein particles from the air. Careful selection of gloves and the management of the air-handling systems can do this. Washing and subsequent meticulous sterilization of latex surgical gloves have been shown effective in the reduction of the incidence of allergic response in Caracas, Venezuela.[1]

There are many groups of people affected by latex allergy, and some of these groups have different routes of entry for the antigenic protein from the latex sap. A summary of each group will be reviewed. The groups being discussed are the general population, patients, and health care workers.

HEALTH CARE WORKERS

Individual studies reported later show a large variation in the incidence of latex allergy from one institution to another. There are several reasons for this variation: the strongest reason is that an institution may use a high antigen content glove and another institution may use a different and lower antigen content glove. Virtually no researchers have addressed this issue. Most institutions use several brands of examination and surgical gloves. As already discussed, the route of entry for health care workers is by inhalation. Exposure can be controlled by adequately cleaning the air to prevent exposure or by eliminating the source. Source elimination can be done based on knowledge of the antigen content of the gloves, and replacement of high antigen content gloves with lower antigen content gloves. In addition, the ventilation systems should be evaluated by an industrial hygienist to ensure efficient fine-particle capture. The evaluation tools are not traditional for this efficiency evaluation. Yunginger and co-workers[2] evaluated 71 lots of typical latex gloves in 1994 and found the allergen levels varied by 3000-fold in these products. He also found that the variation was greater in examination gloves when compared to surgical gloves. A similar study was conducted earlier and published in 1988 by Turjanmaa and colleagues[3] in Finland. The occurrence of latex allergy in Europe preceded the experience in the United States by several years. A total of 19 different gloves were evaluated.

Arellano and co-workers[4] evaluated the prevalence of latex sensitization in physicians in 1992. They contacted surgeons, anesthesiologists, and radiologists employed at a Toronto hospital. All physicians were asked to consent to skin prick testing. Each subject completed a questionnaire concerning demography, atopic history, previous latex reactions, duration and frequency of latex exposure, and outcome from previous surgery. A commercially available skin test reagent was obtained from Bencard Laboratory, Mississauga, Ontario, Canada. (*Note:* Skin testing is not currently permitted by the FDA in the United States because of reported anaphylaxis.) Physicians eligible to participate in the study numbered 179; 11 refused to participate and 67 could not be located on the days of the study. In this study, 9.9% of the physicians were sensitized to latex. Among the latex-exposed physicians, hand sensitivity to latex gloves did not reliably predict latex systemic allergy. Unfortunately, no data were presented concerning the gloves or the aeroallergen concentrations in use during the study.

Questionnaires were distributed to all 1628 U.S. Army dental officers worldwide by Berky and colleagues.[5] Responses were received from 1043 (or 43%) of the dental officers. There were 143 cases (or 13.7%) of latex allergy indicated by the responders. This was a prevalence of 8.8%. No data were offered as to the type of gloves in use or aeroallergen concentrations.

The Alberta Society of Medical Laboratory Technologists membership was invited to participate in a study concerning the sensitivity to latex gloves as reported by Salkie.[6] All participants were sent a questionnaire and asked to supply a 10-ml blood sample for IgE analysis. A total of 230 members participated in the study. Unfortunately, the total number of persons in the society was not identified. However, the results show 122 (or 53%) reacted to latex gloves. There were only three positive IgE results.

Liss and co-workers[7] in 1997 confirmed a higher value for laboratory workers at 16.9% when compared to other study participants. In 1994, Yassin and colleagues[8] investigated the incidence of latex allergy in hospital employees in Ohio. An interview of 224 hospital employees was conducted together with skin prick tests. Of the total subjects evaluated, 136 were nurses, 41 technologists, 13 dental staff, 11 physicians, 6 respiratory therapists, and 17 housekeeping and clerical workers. The study found that 38 individuals (or 17%) were positive for latex sensitization. The authors concluded the incidence of latex allergy was 17% in the hospital studied. No data were offered as to the type of gloves in use or to aeroallergen concentrations. In addition, the total number of employees in each category and the total numbers of hospital workers were not specified.

Hunt and colleagues[9] at the Mayo Clinic in Rochester, MN documented an epidemic of latex allergy. During the time period from January 1990–June 1993 342 employees were evaluated. These employees reported symptoms suggestive of latex allergy. The symptoms were contact urticaria, allergic rhinitis, conjunctivitis, and asthma. Not all people had all symptoms. Persons were interviewed and were skin prick tested. In 12 people 16 episodes of rubber-induced anaphylaxis were documented; 6 occurred after skin testing with extracts from rubber gloves. The anaphylaxis occurrence for these patients prompted researchers to use restraint in skin testing as a diagnostic tool. Of the 342 staff members, 104 were latex allergic, amounting to 30% of the employees evaluated. The peak onset of symptoms occurred in late 1989 and early 1990 and did not correlate with the peak in glove usage. There was no information concerning the type of gloves in use or aeroallergen concentration. The technology was not clear as to air sampling techniques.

An evaluation of the workforce of St. Johns Mercy Medical Center, a 900-bed hospital, was conducted by Kibby and colleagues.[10] A review of purchasing requisitions was conducted. Departments were selected for this study based on both high and low glove use. Even though the study group included selection based on glove use, the study result did not address the issue. There were 135 participants in the study: 11 (or 8.2%) of the sample were skin test positive and 7 of those reported that allergic symptoms developed when they had dermal contact with latex. Data were not supplied concerning the total number of clinical employees.

Sussman and co-workers[11] evaluated the incidence of latex allergy in housekeeping personnel. All 71 members of the housekeeping staff at Wellesley Hospital in Toronto, Canada were invited to participate in the study; 50 volunteered for the study. All wore yellow rubber latex gloves (Cor-Teck) approximately 25–30 h per week. There were four positively latex sensitized housekeeping personnel, for an overall prevalence in this group. The authors indicate this is statistically significant based on the data of the general population (1%) at the time of the study.

Liss and colleagues[7] conducted a large cross-sectional study of health care workers. The study included an evaluation of gloves and aeroallergen concentration, skin prick test, and interviews. All 2062 employees of a general hospital in Ontario, Canada were invited to participate in the study. There were a total of 1351 participants in the study. The mean protein content of the powdered surgical gloves in use was 324 µg/g and of the powdered examination gloves was 198 µg/g. Personal latex aeroallergen concentrations ranged from 5–616 ng/m³ of air. The prevalence of positive latex skin test was 12.1%. Latex skin test positive participants were more likely to have a positive skin test to foods. The prevalence of latex sensitivity was highest among laboratory workers, 16.9%, and nurses and physicians, 13.3%. The highest surgical glove use correlated with the highest skin reactions. The most often reported symptoms among the latex positive people were hives, eye symptoms, and wheezy or whistling chest. This study identifies glove protein content and aeroallergen concentration; and documents a prevalence of 12.1%, which is significant when compared to the incidence in the general population.

GENERAL POPULATION

There are several studies that address the general population. A study was conducted by Ownby (a medical doctor) and co-workers[12] in Michigan. A second study was conducted in the United Kingdom by Merrett and colleagues.[13] Both studies reported measurements of latex-specific antibodies in blood from blood donors. Ownby and colleagues[13] attempted to reduce the number of health care workers in the study so that it represented the general population, and as much as possible excluded health care workers. Ownby excluded blood samples collected from medical clinics in an attempt to measure the level of sensitization in the general population. The Ownby study was conducted in 1995 and published in 1997 in the *Journal of Allergy and Clinical Immunology*. The Merrett study was reported in 1994 at the American College of Allergy, Asthma, and Immunology annual meeting as a progress study. Merrett and co-workers reported on 1006 blood donors and they were continuing to study 5000 blood donors. A reference check in January 1998 did not reveal a newer publication.

The results of the Ownby study are surprisingly high. Blood donor samples collected in Michigan demonstrated 6.4% of the general population had latex antibodies. Ownby excluded Red Cross donor centers located in clinical practice areas as part of the plan to minimize the number of health care workers identified as a high-risk group. Workplace samples were collected over a 3-week period from workplace mobile collection sites. Unfortunately, these locations were not identified. The cause of latex allergy in the general population is not clear. People have contact with many natural rubber-containing articles, such as rubber gloves, condoms, rubber bands, and automotive tires; and each can be a source of latex antigen. Ownby notes that the state of Michigan contains major automotive manufacturing facilities. It is

TABLE 2
Distribution of Positive Test Results for Antilatex IgE Among 1000 Study Subjects by Sex, Age, and Race

| | AlaSTAT | | | | |
	Negative	Positive	Total (%)	p-Value[a]	1990 Census (%)[b]
Sex					
Male	485	45	530 (53)		47.0
Female	451	19	470 (47)	0.014	53.0
Age					
<45 years	665	41	706 (70.6)		
≥45 years	271	23	294 (29.4)	0.14	
Race					
White	829	50	879 (87.9)		77.0
Black	70	10	80 (8.0)		20.6
Other	37	4	41 (4.1)	0.039	2.6

[a] p-Values were determined by using the chi-square statistic.
[b] Data from Census of Population and Housing 1990; Summary Tape File 1 on CD-ROM (Michigan), prepared by the Bureau of the Census, Washington, D.C., 1991.

reasonable to anticipate that exposure to automotive tires is higher to auto industry workers. Unfortunately, countrywide data were not available at the writing of this chapter. The Ownby study indicates more men are sensitized than women (47 men vs. 19 women).

The blood samples in the Ownby study were analyzed by two primary methods, the AlaSTAT procedure (Diagnostic Products, Los Angeles, CA) that currently is the only method approved by the FDA, and the competitive inhibition assay (CAP) (Pharmacia Diagnostics, Dublin, OH). All samples were analyzed by the AlaSTAT procedure, and many by the CAP assay. Not all samples were analyzed by the CAP assay because the volume of blood was not large enough. The AlaSTAT results are presented in Table 2.

The mean age of the blood donors was 37.8 years. Male donor samples were more likely to be positive (8.7% vs. 4.1%). Of the results from the AlaSTAT test, 61% were corroborated by the CAP assay. The AlaSTAT test results demonstrated 6.4% of the population contained the antilatex IgE antibodies, and the CAP test results corroborated 61% of the AlaSTAT test results.

The Merrett and co-workers'[13] study evaluated blood donor samples collected in the United Kingdom. The results demonstrated the overall prevalence rate as 7.7%, with 7.9% for males and 7.4% for females. These results are similar to the Ownby study. Merrett and colleagues[13] planned to study 5000 U.K. blood donors. The reference is for 1006 blood donor samples (544 men and 462 women). Testing was done using the AlaSTAT microplate latex test and Western blotting test procedure. Interestingly, four samples had class III latex IgE antibody levels (all males, less than 30 years old) and were positive in blotting, though not always to the same molecular weight allergens. Generally, health care workers are sensitized to HB5, one of the natural product antigens; and spina bifida patients are sensitized to HB1, another natural product antigen that is also known as the rubber elongation factor. Of the blood donors, 34 were class II (24 males and 15 females).

The results of the Ownby and Merrett studies are surprisingly high because significant exposure is not expected in the general population. Ownby cautions that these antibody concentrations were from a relatively random adult population. Both studies also point out that the clinical significance of these data are not well understood, and that further evaluation is needed. It is also true that the two procedures are relatively new, and comparison with other test methods, such as the radioallergosorbent test (RAST) and enzyme-linked immunosorbent assay (ELISA), is not well defined. In general, RAST results have been thought to underreport or produce false negative results. Early studies suggest the incidence in the general population is 1%, but these results were based on the less well-defined RAST or ELISA procedures.

If these levels of sensitization (Ownby, 6.4%) of the general population are corroborated nationally, then the incidence is significantly high enough to cause concern. These general levels cause us to reevaluate the high-risk groups such as health care workers. The incidence in health care workers in general is similar to the general population. However, the majority of latex sensitized health care workers are women. Most nurses are female and nurses appear to use the majority of gloves. It has been thought that the incidence in health care workers is significantly greater than in the general population. For example, estimates of the incidence of disease published in the early 1990s for health care workers ranged from 3% overall to 6% or higher in surgical units.[14] These earlier test results were based on the then available test procedure such as the RAST procedure. Further work is needed to evaluate the national incidence of latex allergy in the general population and to compare the incidence with data generated about health care workers. The same test methods should be used to estimate the extent of the incidence in the high-risk group in health care.

An increase in asthma has been identified in urban areas. Williams and colleagues[15] selected 50 adult asthmatic patients and measured the IgE (SE) antibodies to latex proteins using the Pharmacia CAP system. They found that in 20% of the cases these patients were latex sensitized; and then went a bit farther with the statistics, extending the cutoff to 10 standard deviations of the negative control and finding that 28% of the samples were positive. They used a glove extract for the test procedure and apparently the concentration of this extract was not calibrated in his work. We know the concentration of antigens in a glove can be as much as 3000 times different from another glove. Therefore, their test results are biased to the control sample used. Their studies (published in 1994), however, indicate that sensitization to latex proteins is highly prevalent in adult asthmatic patients. Latex allergens can be isolated from ambient air, medical, and industrial settings; and hence their contribution to asthma must be seriously considered.

Safadi and co-workers[16] published work in 1995 that indicated immediate hypersensitivity to latex affects 25–50% of children with spina bifida and urogenital birth defects, and 5–15% of all health care workers. They also indicated that latex products utilized by individuals in tire manufacturing and by housekeepers create exposure for those workers. They further stated in their article that about 17 million people presently in the general population are latex sensitized, an estimate of 1% of the population. The Ownby and other studies suggested that this number is far higher. The Safadi study reported anaphylaxis secondary to latex exposure during routine medical examination. Both exposures occurred to individuals who had only minor local reactions following latex exposure with no prior history of anaphylaxis. The first incident involved a 34-year-old medical secretary who developed generalized hives and hypotension minutes following a routine pelvic exam with latex gloves. The second incident involved a 60-year-old emergency care physician who underwent a routine rectal examination with latex gloves and developed respiratory distress and circulatory collapse minutes later. These incidents point to the very seriousness of anaphylactic reactions with only minor exposure to latex-containing gloves. The work, however, also points to the fact that tire manufacturing workers are at risk.

Irregular black objects, which have been observed in urban air samples, have been identified as tire particles. These particles contain latex allergens. In 1995, Hornberger and Portnoy[17] began measurement of tire particles in the urban air of Kansas City, MO. Samples were collected continuously for 24 h daily with a Burkhard trap. Particles were counted every 2 h using an optical microscope at 400 power. Variations occurred depending on the time of day. The highest concentration of particles occurred in mid-morning (5564 per cubic meter), and in the evening (5232 per cubic meter). The lowest counts occurred in late afternoon (2064 per cubic meter) and early morning (2916 per cubic meter). These correlated with the peak traffic flow, coming after the morning and the evening rush hours. The evening count peaked between 8 PM and midnight, suggesting the particles remained disbursed in the air for a long period of time. The median counts were apparently unrelated to daily temperature fluctuations.

Hornberger and Portnoy concluded that tire particles constitute a substantial part of urban airborne pollution. Since these rubber particles contain the latex allergen, they pose a potential health risk. The late evening peak in tire particles may be associated with the worsening of asthma either indirectly or directly due to sensitivity to latex. They suggested that further monitoring of these exposures is critical to understanding the effect of airborne tire particles on respiratory disease.

In January of 1995, Williams and colleagues[18] launched an interesting study connected with respirable particulate air pollution reports. They reported that low concentrations of respirable particulate matter have been significantly associated with the increase of daily mortality rates, acute bronchitis, and hospital visits for treatment of lower respiratory tract symptoms. Morbidity and mortality rates for respiratory asthma have increased dramatically, particularly in urban areas, throughout the world. Children seem to be at a particularly high risk and particulate air pollution at the 2–10-μm range has been implicated.

Particulate air pollution in many urban areas consists of small or irregular shaped black particles, in addition to normal particulates. The most likely source of these particles is the abrasion of rubber tires on road surface. Tires are manufactured of natural and synthetic rubber, along with sulfur, carbon black, and zinc oxide. The methods used by Williams and co-workers[18] were for a particular sample collection by a rotary impact sampler, a high-volume sampler, and the passive gravametric air sampler with glass plates. The impaction sampler was 7.4 m aboveground level, and 48 m from a four-lane, moderately traveled road way in Colorado. Additional sampling stations in the metropolitan Denver area were also used. The size of the particulates was determined by optical microscope at a magnification of approximately 450. The microscope was fitted with a glass reticle to permit measurement of particulate sizes. Particle counts were performed by counting 1000 consecutive particles on five separate samples (n = 5000). Identification of the particulates was conducted by both solubility tests and gas chromatographic mass spectroanalysis to confirm chemical composition. In addition, proteins were extracted from the environmental samples and from several standards including automobile tires and gloves, and then analyzed within the CAP system. The results of the study demonstrated airborne quantities of particulates ranging from 3800–6900 per cubic meter.[3] Their size varied greatly, but over 90% of the particulates were between 2.2 and 35.2 μm range. Close to 60% of the particulates were in the respirable dust range of less than 10 μm. There was a sharp drop-off in the number of particulates below 4 μm. The serological results demonstrated a high concentration of latex antigens in the rubber samples. All but one of the sera containing latex-specific IgE were inhibited by the rubber tire extract, with varying degrees of inhibition. For each serum sample, the inhibition was generally equivalent to that achieved with the latex glove extract (1 mg of protein per milliliter) but occasionally higher with the rubber fragment extract. The use of natural rubber latex as a constituent of most tires presents the possibility the tire fragments could be immunologically active. This possibility was verified by demonstrating elutable latex antigen activity by the inhibition assays.

In 1981, the Environmental Protection Agency (EPA) laboratory characterized tire wear particulates with the use of a stainless steel wearing surface. This study failed to demonstrate rubber particulates in the 1–10 μm range.

Rubber tires are manufactured with varying concentrations of natural latex and synthetic rubber. Truck tires have the highest amount of natural latex. Natural latex is desirable because it causes less friction and can withstand higher temperatures, but it has the propensity to release smaller particulates on abrasive wear. Radial tires, as opposed to bias-ply tires, constitute the release of smaller particulates. The increased prevalence of radial tire use over the past two decades could account for the smaller particulates seen in this study.

Miguel and colleagues[19] collected ambient samples in California at two sites in the Los Angeles basin. They both were collected in 1993. Samples were extracted and utilized by a modified-ELISA inhibition assay and a Western blot analysis that were used to identify latex allergens. The results revealed tire tread sources in ambient freeway dust, as well as latex sap and latex glove extracts, to be significant. Levels of extractable latex antigen per protein extracted were about two orders of magnitude lower for tire tread as compared to latex gloves. The Western blot analyses, using binding of human IgE from latex-sensitized patients, caused a band at 34–36 kDa in all tire and ambient samples. Long Beach and Los Angeles, CA air samples showed four additional bands between 50 and 135 kDa.

PATIENTS

There is a concern that the patient entering a health care facility, such as a hospital, nursing home, physician's office, or dental office, may be latex sensitized or become latex sensitized and respond to latex products used in health care. This is not a new problem in patient care. There are some reports that date to the 1930s and refer to allergic reactions to the latex proteins. Wilson[20] reported in 1960 that other researchers believed the rubber itself was the cause of allergic dermatitis rather than the accelerators, which have been very well defined.

There have been frequent reports of anaphylaxis, and some that resulted in death were reported by the FDA in 1991. The preceding section indicated the incidence of latex sensitization in the general population is close to 10%. Therefore, the probability of a patient entering a health care facility being latex sensitized is also close to 10%. In addition, previous medical intervention such as multiple surgeries place some patients in a high-risk category. This requires that the health care provider be aware that the patient may be at risk, and therefore has available appropriate protection or intervention so that the latex-sensitized patient does not sustain a reaction. In a way, the best place for an anaphylactic reaction to occur is in a hospital and perhaps on an operating table. The anesthesiologists are trained to respond to any emergency, can recognize anaphylaxis, and take appropriate measures to intervene. Often, epinephrine is given to the patient to prevent further reaction.

The casual observer and some not so casual observers suggest the best solution to prevent a latex allergic response is a latex-free environment. This is virtually impossible to obtain. Latex is ubiquitous. Latex is in many items, such as paper towels, syringe plungers, surgical masks, and elastic in underwear. The cause of an allergic response is a specific group of proteins. These proteins are water soluble, and if an item is washed, the water-soluble proteins can be removed. Capriles-Hulett and co-workers[1] reported that only 4.3% of the patients tested in Venezuela were found to be latex sensitized and suggest that one reason for this is the washing and resterilization of surgical gloves.

Pearson and colleagues[21] published a study in 1994 of an evaluation of 119 families with children with spina bifida. The work was conducted at the Arnold Palmer Hospital for Children and Women in central Florida. The researchers compared the potential risk factors for developing latex allergy between children with and without latex allergy. The study was conducted by the use of questionnaires. Latex allergy was defined as the development of rhinitis or conjunctivitis, sneezing, hives, angioedema, dyspnea, flushing, wheezing, dizziness, or anaphylaxis in association with the use of latex products. Information collected included gender; race and ethnicity; and history of asthma and other allergic disease; and the number of procedures requiring general anesthesia. Data were also collected on the history and duration of clean intermittent catheterization (CIC), history of constipation and manual evacuation, glove use during CIC and manual evacuation, and history of bladder stimulation procedure. Specific latex products were reviewed when symptoms occurred.

Responses were received from 110 family members. The children were predominately white, non-Hispanic (76%), and mostly male (52%); and 35% had a known history of allergy, such as to drugs or medications. Hay fever occurrence was 13% and asthma was 6%. Of these children, 12% were latex allergic.

Most children developed latex allergy after exposure to balloons (92%). Other symptoms resulted from exposure to condoms, dental dams, rubber-soled shoes, or urinary catheters. The oldest child was 22 years of age. The most frequently reported symptoms were rhinitis or conjunctivitis (92%), angioedema (83%), urticaria (69%), sneezing (67%), dyspnea (54%), wheezing (23%), and dizziness (17%). Surprisingly, children without any allergic history were more likely to become allergic to latex. In addition, latex allergy was more likely in patients who had surgical procedures for which general anesthesia had been administered. Children with latex allergy were more likely to have been using CIC for greater then 18 months.

In other studies, such as that conducted by Ellsworth and co-workers,[22] the incidence of latex allergy in spina bifida patients is much higher. They studied 50 patients and identified 60% as being allergic to latex. Other researchers have reported similar high levels, such as Slater and colleagues[23] who reported 41% in 1990. There is a successful support group for parents of latex-sensitized children that can be reviewed on the Internet.

Moragues and co-workers[24] evaluated 100 children with a mean age of 7.5 +/- 4.8 years diagnosed with myelomeningocele (MMC). Latex allergy was determined by skin test and latex-specific IgE. Of the 29% that were sensitized, 15 were symptomatic and 14 were asymptomatic. Neither sex nor family background showed statistical significance. The authors concluded that the incidence of latex allergy for these MMC patients is related to multiple antigen contact and the operations undergone through a patient's experience. These authors also suggest all patient care should be in a latex-free environment.

Lebenbom-Mansour and colleagues[25] conducted a study of 996 ambulatory surgical patients preoperatively, and published the findings in 1997. A questionnaire addressing demographics, previous surgeries, history of atopy, previous exposure or reactions to latex, congenital abnormalities, and food allergies was completed by all participants. Serum antilatex IgE levels via the AlaSTAT test were compared to the questionnaire results, 6.7% of which had IgE antibodies against latex. Male gender, non-Caucasian race, age, asthma, spinal cord abnormalities, food allergies, stated latex allergy, and symptoms when exposed to latex increased the risk of latex sensitivity. The predictive value of history was low. This study indicates latex allergy to be a significant potential problem for ambulatory surgical patients.

The Capriles-Hulett and co-workers'[1] study conducted in Venezuela offers an interesting number of conclusions, and an insight into medical care in another part of the world. The study indicates 4.3% of the spina bifida patients are sensitized to latex. This is a very low incidence when compared to spina bifida patient incidence in the United States. They attribute this relatively low incidence to lower exposure to latex products. The exposure is reduced in Venezuela because nonlatex bladder catheters are used; the frequency of operation per patient is low; and surgeons' gloves are washed, sterilized, and reused.

The study of latex allergy is something like reading a book. We read a chapter, and think we understand, and perhaps wonder about other relationships. We are in only the first few chapters of understanding latex, and I am sure the following chapters will continue to be exciting and rewarding.

REFERENCES

1. Capriles-Hulett, A., Sanchez-Borges, M., Von-Scanzoni, C., and Medina, J.R., Very low frequency of latex and fruit allergy in patients with spina bifida from Venezuela: influence of socioeconomic factors. *Ann. Allergy Asthma Immunol.,* 75(1):62–64, 1995.
2. Yunginger, J.W., Jones, R.T., Franswat, A.F., Kelso, J.M., Warnes, M.A., and Hunt, L.W., Extractable latex antigens and proteins in disposable medical gloves and other rubber products, *J. Allergy Clin. Immunol.,* 93(5):836–842, 1994.
3. Turjanmaa, K. Laurila, K. Maikinen-Kiljunen, S., and Reunala, T., Rubber contact urticaria, allergenic properties of 19 brands of latex gloves, *Contact Dermatitis,* 19:362–367, 1988.
4. Arellano, R., Bradley, J., and Sussman, G., Prevalence of latex sensitization among hospital physicians occupationally exposed to latex gloves, *Anesthesiology,* 77:905–908, 1992.
5. Berkey, Z., Lucaino, J., and James, W.D., Latex glove allergy, *JAMA,* 268(19): 2695–2697, 1992.
6. Salkie, M.L., The prevalence of atopy and hypersensitivity to latex in medical laboratory technologists, *Arch. Pathol. Lab. Med.,* 117:897–899, 1993.
7. Liss, G.M., Sussman, G.L., Deal, K., Brown, S., Ciovidino, M., Siu, S., Beezhold, D.H., Smith, G., Swanson, M.C., Yunginger, J., Douglas, A., Holness, D.L., and Lebert, P., Latex allergy: Epidemiological study of 1351 hospital workers, *Occup. Environ. Med.,* 54(5):335–342, 1997.
8. Yassin, M.S., Lierl, M.B., Fisher, T.J., O'Brien, K., Cross, J., and Steinmetz, C., Latex allergy in hospital employees, *Ann. Allergy,* 72(3):245–249, 1994.
9. Hunt, L.W., Fransway, A.F., Reed, C.E., Miller, L.K., Jones, R.T., Swanson, M.C., and Yunginger, J.W., An epidemic of occupational allergy to latex involving health care workers, *J. Occup. Environ. Med.,* 37(10):1204–1209, 1995.
10. Kibby, T., and Aki, M., Prevalence if latex sensitization in a hospital employee population, *Ann. Allergy Asthma Immunol.,* 78:41–44, 1997.
11. Sussman, G.L., Lem, D., Liss, G., and Beezhold, D., Latex allergy in housekeeping personnel, *Ann. Allergy Asthma Immunol.,* 74(5):415–418, 1995.
12. Ownby, D.R., Ownby, H.E., McCullough, J., and Shafer, A.W., The prevalence of anti-latex IgE antibodies in 1000 volunteer blood donors. *J. Allergy Clin. Immunol.,* 97(6):1188–1192, 1996.
13. Merrett, T.G., Merrett, J., and Kekwick, R., Prevalence of latex specific IgE in the UK, ACAAI Meeting Abstracts, 1994.
14. Gonzales, E., Latex hypersensitivity: A new and unexpected problem. *Hosp. Pract.,* February 15:137–151, 1992.
15. Williams, P.B., Garcia, M.D., and Selner, J.C., Prevalence of IgE anti-latex in asthmatics. Meeting Abstracts, Denver, CO, 1994.
16. Safadi, G.S., Wagner, W.O., Pien, L.C., and Melton, A.L., Latex-induced anaphylaxis following routine medical examination. Meeting Abstracts, Cleveland, OH, 1995.

17. Hornberger, B., and Portnoy, J., Measurement of tire particles in urban air, Meeting Abstracts, Kansas City, MO, 1995.
18. Williams, P.B., Buhr, M.P., Weber, R.W., Volz, M.A., Kopeke, J.W., and Selner, J.C., Latex allergen in respirable particulate air pollution, *J. Allergy Clin. Immunol.*, 95(1):88–94, 1995.
19. Miguel, A.G., Glovsky, M.M., Weiss, J., and Gass, G.R., Latex allergens in tire dust and airborne particles, Pasadena, CA, *Environ. Health Perspect.*, Nov. 104(11):1180–1186, 1996.
20. Wilson, T.H., Rubber-glove dermatitis, *Br. Med. J.*, July 2: 1960.
21. Pearson, L.P., Cole, J.S., and Jarvis, W.R., How common is latex allergy? A survey of children with myelodysplasia, *Dev. Med. Child Neurol.*, 36:64–69, 1994.
22. Ellsworth, P.I., Merguerian, P.A., Klein, R.B., and Rozycki, A.A., Evaluation and risk factors of latex allergy in spina bifida patients: Is it preventable?, *J. Urol.*, 150: 691–693, 1993.
23. Slater, J.E., Mostrello, L.A., Shaer, C., and Honsinger, R.W., Type I hypersensitivity to rubber, *Ann. Allergy*, 65:411, 1990.
24. Moragues, E.F., Garcia-Ibarra, F., Hinarejos, D.C., Verduch, M., Ruiz, R.C., Ramos, M.A., and Garcia N.A., Latex allergy in children with myelomeningocele. Incidence and associated factors, *J. Rheumatol.*, 24(9):1826–1829, 1997.
25. Lebenbom-Mansour, M.H., Zaglaniczy, K., Post, A.K., Jennett, M.K., Ownby, D.R., and Oesterle, J.R., The incidence of latex sensitivity in ambulatory surgical patients: A correlation of historical factors with positive serum immunoglobin E levels, *Anesth. Analg.*, 85(1):44–49, 1997.

8A Violence in the Workplace: A Growing Crisis Among Health Care Workers

Jane Lipscomb

CONTENTS

BACKGROUND

Violence toward health care workers has a remarkable history that includes an 1849 report of the first documented case of a patient fatally assaulting a psychiatrist. Numerous reports of fatal and nonfatal assault toward psychiatrists and other health care workers have been noted in the mental health and biomedical literature since that time. However, it was not until the late 1980s that workplace violence began to be recognized as an occupational hazard. In 1990, workplace violence bypassed machine-related deaths to become the second leading cause of occupational traumatic injury death overall. By 1993, the California Occupational Safety and Health Administration (OSHA) had developed the first set of guidelines to highlight the problem of violence in the health care and service sectors. These guidelines entitled "Guidelines for Security and Safety of Health Care and Community Service Workers" were a model for future state and federal activities in the area.

In spite of the magnitude of the problem and the recognition of the need for guidelines in the area, it was not until 1993 that the federal OSHA issued its first citation against an institution, Charter Barclay Hospital, for violent assaults toward workers under the Occupational Safety and Health (OSH) Act, 5(a)(1) "general duty clause." OSHA later followed California's lead with its 1996 "Guidelines for Preventing Workplace Violence Among Health Care and Social Service Workers".

EXTENT OF THE PROBLEM

According to the Bureau of Labor Statistics (BLS, 1994), 48% of all reported cases of nonfatal workplace assaults are committed by health care patients. In addition, two thirds of all nonfatal assaults occurred in nursing homes, hospitals, and establishments providing residential care and other social services. The rate of nonfatal assaults among "nursing and personal care facilities" workers was 38/10,000 compared with a rate of 3/10,000 in the private sector, a nearly 13-fold difference. Interestingly, 30% of the victims of these assaults were government employees even though they comprise only 18% of the workforce. A study conducted by Love and Hunter (1996) demonstrated that there was a rate of 14.6–32.0 per 100 full-time equivalent (FTE) OSHA recordable injuries for workplace assaults among nursing staff in five public sector psychiatric facilities. These rates are known to be very conservative given the strict OSHA definition of "recordable injury," which specifies that the injury must result in loss time, loss of consciousness, restricted work, or medical treatment other than first aid. This rate compares with a rate of 8.8 per 100 FTE for all private sector workers, 8.3 per 100 FTE for mining and 14.2 per 100 FTE for heavy construction. An earlier report (Wasserberger, 1989) reported that weapon carrying in hospital emergency departments was pervasive with 25% of major trauma patients carrying weapons in one urban hospital.

Between 1980 and 1990, 106 health care workers were murdered on the job. Among them were 27 pharmacists, 26 physicians, 18 registered nurses (RNs), and 7 nurses' aides. Of the victims, 18 were classified as other health care workers (Goodman et al., 1994).

In addition to suffering physical assault on the job, health care workers, nurses specifically, suffer from workplace sexual harassment. According to Williams (1996), 64% (n = 224) of nurses surveyed reported personal experience with sexual harassment. Of these, 73% reported patients or client perpetrators and 58% reported physician perpetrators.

PREDISPOSING FACTORS FOR EPIDEMIC OF VIOLENCE IN THE HEALTH CARE SETTING

A number of risk factors have been identified as contributing to the epidemic of workplace violence in the health care setting including the prevalence of handguns and other weapons in society, the crisis in the mental health care system that has resulted in a lack of treatment for those in need, the availability of drugs and money in hospitals, and the unrestricted movement within some clinics and hospital. Environmental factors that are thought to contribute to the problem include poor security, inadequate training, staffing levels and patterns, time of high activity, containment activities, and long waits in emergency departments.

Hospital and service sector work site safety and health programs must take into account the risk of workplace assault. A comprehensive program must contain the following four elements:

1. Management commitment and employee involvement
2. Work site analysis
3. Hazard prevention and control
4. Safety and health training

Engineering controls include security guards, metal detectors, access control, bullet-proof glass and partitions, silent alarms, two-way radios, escort services, lighting and other environmental factors, and changes in office design to provide an escape route for workers. Administrative controls include work practices such as acuity-based staffing patterns, banning of working alone after dark, minimization of cash and drugs on hand, development of perpetrator profiles (e.g., mentally ill, gang members, drug use, history of violence), and clear policies and procedures. Education and training must be part of a comprehensive program and should focus on conflict resolution and nonviolent response, security, and rights and liabilities. In addition, a program should include a performance-based evaluation of training, and a strategy for training that provides coverage for the employees involved. Education concerning safety on the job must begin during vocational and professional training.

We know much about the magnitude and risk factors for workplace assault in health care. We must now develop strategies to measure the impact of OSHA and local guidelines while at the same time demanding greater protection from government and hospitals, including workplace violence in collective bargaining.

REFERENCES

Bureau of Labor Statistics (BLS). *Work injuries and illnesses by selected characteristics, 1992.* U.S. Department of Labor, Bureau of Labor Statistics, 1994; Washington, D.C.

Goodman, R., Jenkins, L., and Mercy, J. Workplace-related homicide among health care workers in the United States, 1980 through 1990. *JAMA,* 1994; 272(21): 1686–1688.

Love, C., and Hunter, M. Violence in public sector psychiatric hospitals. *Journal of Psychosocial Nursing,* 1996; 34(5):30-3473–30-3477.

OSHA. *Guidelines for preventing workplace violence for health care and social service workers.* U.S. Department of Labor, Occupational Safety and Health Administration, 1996; OSHA 3148.

State of California. *Guidelines for security and safety of health care and community service workers.* Sacramento, CA: Division of Occupational Safety and Health, Department of Industrial Relations, 1993.

Wasserberger, J., Ordog, G., Kolodny, M., and Allen, K. Violence in a community emergency. *Archives of Emergency Medicine,* 1989; 6:266–269.

Williams, M. Violence and sexual harassment: Impact on registered nurses in the workplace. *AAOHN.* 1996; 44(2):73–77.

8B Violence in Washington State Health Care Workplaces, 1992–1995*

Michael Foley, Barbara Silverstein, and John Kalat

CONTENTS

INTRODUCTION

Health care workers, particularly those in psychiatric facilities, are at higher risk of physical assault than those in other occupations. Over the study period, 35% of all assault-related workers' compensation claims occurred in the health services industries. Moreover, health-related occupations accounted for approximately 46% of all assault-related claims over the same period. Social service workers are also at higher risk of assault because of their frequent contact with a distressed or impaired population. While the trend is falling over this period, assault rates in the health care sector remain far higher than those in other service industries.

Violence in the workplace is receiving increasing attention from both the media and the public health community. Although the media tends to be drawn to episodes of spectacular impact, such as homicides perpetrated by disgruntled former employees or co-workers, public health research is focusing on those incidents where the assailant is either unknown to the victim or is under the victim's care as either a patient or a recipient of social services. The reasons for this focus are because the cases where the assailant is either a personal acquaintance or former co-worker of the victim are relatively rare, and because these incidents appear to be less predictable and therefore less amenable to safety intervention (Toscano, 1996).

---

* This chapter reports on evidence drawn from a study of Washington State's workers' compensation to describe trends in workplace violence for the period from 1992 to 1995.

0-8493-3382-2/99/$0.00+$.50
© 1999 by CRC Press LLC

Workplace violence is not evenly distributed across all industries and occupations. Some occupations and industries are at higher risk because they involve the exchange of money, face-to-face transactions with the public, and working alone or at night. Such industries as convenience stores, gas stations, hotels, and taxicab services are representative of such risks where the motivation of the assailant is frequently robbery (Castillo and Jenkins, 1994).

Other industries are at higher risk for violence because they may combine the preceding risks with work involving a distressed or constrained population. Health care workers, especially those in psychiatric facilities, are at higher risk of physical assault, even though such attacks may not fall under the legal definition of criminal acts. Social service workers are also at higher risk of assault because of their more frequent contact with a distressed or impaired population (NIOSH, 1996). In all these cases, there is evidence of predictable risk factors that increase the likelihood of assault. This counters the notion that violence on the job is a random event, and consequently impervious to remedy.

The figures in this study are based on workers' compensation claims related to assaults and violence collected by Washington State Department of Labor and Industries (L&I). The objectives are to identify occupational and industrial groups at elevated risk of workplace violence and assault, and to describe trends in the pattern of risk.

METHODS

Data for workers' compensation claims related to assaults and violence for the study period were obtained from L&I files. In Washington, employers are required to obtain workers' compensation insurance through the department's industrial insurance system unless they are qualified to self-insure. Self-insurance is permitted if a firm is able to set aside sufficient reserves and meet certain guidelines; roughly 400 (chiefly large employers) currently do so.

Approximately two thirds of Washington state workers are covered through the state fund. Excluded from the system are workers covered by another insurance system (e.g., federal employees) and a few select groups (e.g., corporate officers, domestic employees) for whom coverage is optional. The department maintains files for both state fund and self-insured employers, although the information collected from self-insured companies is more limited, which will be explained later.

The Department of Labor and Industries defines claims either as "non-compensable" (for which injured employees are reimbursed for medical treatment costs only) or "compensable" (for which both medical costs and wage-replacement benefits for lost workdays are paid). To qualify for definition as a "compensable" claim, the injury must have resulted in four or more lost workdays.

Claims for workers' compensation are coded by industry and occupation of the claimant; injury source, nature, body part, and type of event or exposure causing the injury are coded using U.S. Department of Labor Z16.2 codes, developed by the American National Standards Institute. The coding system does not specifically designate injuries as assault- or violence-related, so criteria were established by the authors to define cases as assault related. These criteria are summarized in the Appendix at the end of this chapter.

One advantage of using the state's workers' compensation data to estimate the number and incidence rate of assaults is that all claims are coded, not only those resulting in days lost from work, as in the Bureau of Labor and Statistics (BLS) survey of occupational injuries and illnesses. This allows the tracking of a far greater proportion of assaults than is possible with the BLS survey. Another advantage is that since the workers' compensation data are not sample-based estimates, as in the BLS survey, the problems of small samples are avoided. The disadvantage of using workers' compensation data to estimate rates of workplace violence is that it is known that only a small fraction of all cases of workplace violence result in a claim for workers' compensation. If the assault does not result in an injury requiring medical treatment or lost workdays, then it does not appear as a case. In a study of assaults against staff at two Washington State psychiatric hospitals it was estimated that less than 5% of all cases of assault resulted in a claim for workers' compensation (Bensley et al., 1997).

The state fund database includes claims involving medical treatment or time lost from work; the self-insurance database includes only claims involving 4 days or more of time loss (with some exceptions). Information on employment is reported to the department by state fund employers, including the number of hours worked by employees. This allows the calculation of incidence rates, defined in this report as number of claims per 10,000 full-time equivalent (FTE) workers per year. (Full-time equivalency assumes that each full-time employee works 2000 hours/year.) Work hours are not collected for self-insured companies; thus it is not possible to calculate claim rates for these businesses.

To observe trends over time within industrial and occupational categories, claims data are presented for all years from 1992–1995, as well as averages for the entire period.

RESULTS

There were an average of 2529 claims related to assaults and violence each year in Washington State from 1992–1995 (Table 1). Over this period there was a decrease of approximately 12% in the number of such claims. Of the total number of claims, 666 were "compensable," that is, involved four or more lost workdays. For the entire period, total claim costs averaged almost $9 million per year. For the state fund companies the average cost per claim was roughly $3,500, with compensable claims averaging over $14,500 each and noncompensable (medical only) claims averaging almost $600.

In both the self-insured and state fund categories the percentage of claimants who are female was approximately 58%. The average age of the claimants was 34.7 for the state fund, and slightly older (37.9) for the self-insured employers. The most frequent type of injury was "struck or beaten by fellow worker, patient, etc.," reflecting the large number of assault-related injuries to health and social service workers.

TABLE 1
Major Industries with the Largest *Numbers* of Workers' Compensation Claims Related to Assaults and Violence, Washington State, 1992–1995

SIC[a]	Description	1992 No.	1993 No.	1994 No.	1995 No.	Average 1992–1995 No.
80	Health services	885	956	879	768	847
83	Social services	554	528	496	407	496
58	Eating and drinking places	159	147	131	131	142
82	Educational services	113	126	127	126	123
91	Public administration: executive, legislative and general	126	109	115	115	116
92	Public administration: justice, order and safety	83	77	68	64	73
70	Hotels and other lodging places	55	70	72	65	66
73	Business services	78	69	58	59	66
94	Public administration: human resource programs	28	35	74	69	52
	Total (all SIC codes combined)	2647	2635	2511	2320	2529

Note: Numbers are calculated using both time-loss and medical-only cases. Assault- and violence-related claims were defined as those with Z16.2 type of event codes 023, 025, 026, 027, and 502; and any source codes except 0200, 0201, 0230, 0240, 0250, 0270, and 5910.

[a] SIC, Standard Industry Code.

The industries with the largest numbers of assault- and violence-related claims are shown in Table 1. Health services, social services, eating and drinking places, and public administration (executive, legislative, and general) continued to lead the list of industries with the highest number of assault-related claims. There was a drop in claims numbers in all four of these sectors over the period, which accounted for nearly all the overall decrease in claims. Educational services showed an increase in claims over the period and, notably, public administration—human resource programs showed an increase of 146% in claim frequency, making it the sixth leading sector in 1995. For the entire period, however, over half of all injuries occurred in health and social services.

Claims rates were calculated for all state fund employers by the industrial sector. These are presented in Table 2. The health care and social services sectors experienced assault-related claims rates on average six to seven times greater than those for all industries combined. More striking still are the rates for the most risky subsectors within health care. Where skilled nursing care (Standard Industry Code [SIC] 8051) experienced an assault claims rate 15 times that of the typical industry, that for psychiatric hospitals (SIC 8063) averaged 908 claims per 10,000 full-time equivalents (FTEs), a rate almost 48 times as high as the typical industry. This is shown graphically in Figure 1.

TABLE 2
Industries with the Highest Workers' Compensation Claims *Rates,* State Fund Employers, Washington State, 1992–1995

	Workers Compensation Claims Rate per 10,000 Workers				
Industry	1992	1993	1994	1995	Average 1992–1995
1. Psychiatric hospitals (SIC 8063)	842	893	1016	880	908
2. Residential care (SIC 8361)	573	544	501	338	489
3. Skilled nursing care (SIC 8051)	286	324	286	256	288
4. Nursing/personal care (SIC 8059)	236	201	221	164	206
5. Job training services (SIC 8331)	113	101	140	170	131
6. Detective and armored car services (SIC 7381)	145	120	112	101	120
7. Police protection (SIC 9221)	136	58	99	65	90
8. Administration of social and manpower programs (SIC 9441)	57	63	121	110	88
9. Correctional institutions (SIC 9223)	68	100	60	62	73
10. Rooming and boarding houses (SIC 7021)	71	65	72	59	67
11. General medical/surgical hospitals (SIC 8062)	66	72	73	54	66
12. Child day care services (SIC 8351)	72	49	56	60	59
13. Specialty outpatient clinics (SIC 8093)	42	57	73	62	59
14. Membership organizations, n.e.c. (SIC 8699)	46[a]	63	28[a]	68	51
15. Drinking places (SIC 5813)	61	37	42	53	48
16. General government, n.e.c. (SIC 9199)	56	50	40	40	47
17. Home health care services (SIC 8082)	42	24[a]	62	53	45
18. Elementary and secondary schools (SIC 8211)	35	44	36	39	39
19. Apartment building operators (SIC 6513)	35	36	27	36	34
Weighted average for all industries	20.5	20.3	18.5	16.7	19.0

Note: The following detailed industries had more than 10 state fund injury claims related to workplace violence, 100 or more full-time employees per year, and at least twice the 1995 average overall rate of 17 such claims per 10,000 workers. All claims are included, whether or not they involve time loss. Assaults were defined as incidents coded as type 23, 25, 26, 27, and 502 (respectively "kicked by;" "bit by;" "struck or beaten by fellow worker, patient, etc.;" "struck or beaten by person in the act of a crime;" and "shot by another person"), which were included as before. Type 28 ("stabbed by") was included only when the source was another person. In addition, all sources except animals or the claimant were included. Note that all claims are included, whether or not they resulted in time loss.

[a] Fewer than ten claims.

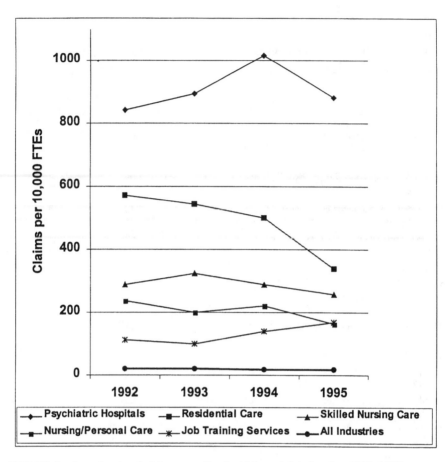

FIGURE 1 Assault-related claims rates for the highest risk industries, 1992–1995.

Across all industries there was a decrease over the period from 20.5 claims per 10,000 FTEs to 16.7. When limited only to compensable, time-loss claims, the average overall rate fell from 4.4 claims per 10,000 FTEs to 3.4. Most of the decrease was driven by declines in claims rates for the riskiest industries. Claims rates fell for social services, health services, and all public administration sectors with the conspicuous exception of human resource programs. Among the top three industries there was no change of rank between 1992 and 1995. The industries showing the most significant increase in claims rate were administration of social and manpower programs (SIC 9441) rising from 13th to 6th place, and job training services (SIC 8331), which rose from the 7th to the 4th highest claims rate over the period.

For the health care sector as a whole, the assault-related claims rate fell from 120 in 1992 to 101 in 1995, but this summary figure masks wide differences within subsectors of the health care industry in Washington. Generally speaking, the greatest decline in rates occurred in the subsectors with the highest initial rates of violence, including both the skilled nursing care (SIC 8051) and the nursing/personal care (SIC 8059) sectors. The former includes convalescent homes with continuous nursing care,

TABLE 3
Workers' Compensation Claims *Rates*, State Fund Employers in the Health Services Sector, Washington State, 1992–1995

	Workers Compensation Assault Claims Rate per 10,000 Workers								
Industry	1992		1993		1994		1995		Average 1992–1995
	Rate	(No.)	Rate	(No.)	Rate	(No.)	Rate	(No.)	Rate
1. Psychiatric hospitals (SIC 8063)	842	(249)	893	(257)	1016	(272)	880	(230)	908
2. Skilled nursing care (SIC 8051)	286	(496)	324	(564)	286	(452)	256	(394)	288
3. Nursing/personal care (SIC 8059)	236	(38)	201	(38)	221	(42)	164	(33)	206
4. General medical hospitals (SIC 8062)	66	(13)	72	(14)	73	(13)	62	(21)	66
5. Home health care services (SIC 8082)	42	(13)	24	(8)	62	(13)	54	(12)	45
6. Specialty outpatient clinics (SIC 8093)	42	(13)	57	(19)	73	(26)	53	(21)	59
All industries	21	(2473)	20	(2474)	19	(2343)	17	(2132)	19

extended care facilities, and mental retardation hospitals. The latter category includes nursing homes without continuous nursing care, personal care homes with health care, and convalescent homes for psychiatric patients. Psychiatric hospitals (SIC 8063), as a category, was conspicuous among the highest risk sectors for its lack of any discernible downward trend over the period, although its rate has declined from its 1994 peak, when more than one assault-related claim was filed for every ten workers in the industry. Sectors of the health care industry for which initial 1992 claims rates were low experienced climbing rates over the study period. This was the case for both specialty outpatient clinics (SIC 8093) and home health care services (SIC 8082).

To examine the trend in severe injuries due to assault in the health care industry, we extracted time-loss claims (called "compensable") from the total. These are presented in Table 4 together with the claims rates and frequencies for both medical-only and time-loss claims in Table 3 for the six highest risk health care industries for 1992 through 1995. The rate for all industries is shown for purposes of comparison. For total claims, as we have seen, the rate declined for skilled nursing care, nursing and personal care, and general medical and surgical hospitals. The rate did not fall for psychiatric hospitals, home health care services, or specialty outpatient clinics. However, when considering only the more serious claims, the time-loss claims, the rate fell only for skilled nursing care and for nursing and personal care. The rate rose substantially for psychiatric hospitals, general medical hospitals, and home health care services.

Occupations with the largest numbers of claims related to assaults and violence are shown in Table 5. Unsurprisingly, nursing aides and orderlies, health aides, health

TABLE 4
Workers' Compensation Claims *Rates*, State Fund Employers in the Health Services Sector, Washington State, 1992–1995 *Time-Loss Claims Only*

Industry	Workers Compensation Assault Claims Rate per 10,000 Workers								Average 1992–1995
	1992		1993		1994		1995		
	Rate	(No.)	Rate	(No.)	Rate	(No.)	Rate	(No.)	Rate
1. Psychiatric hospitals (SIC 8063)	281	(83)	257	(74)	388	(104)	371	(97)	324
2. Skilled nursing care (SIC 8051)	54	(93)	51	(89)	41	(64)	29	(44)	44
3. Nursing/personal care (SIC 8059)	50	(8)	42	(8)	37	(7)	25	(5)	39
4. General medical hospitals (SIC 8062)	6	(2)	6	(2)	17	(3)	18	(4)	12
5. Home health care services (SIC 8082)	5	(1)	6	(2)	28	(10)	10	(4)	12
6. Specialty outpatient clinics (SIC 8093)	3	(1)	5	(1)	8	(3)	9	(3)	6
All industries	4.4	(534)	4.2	(508)	4.3	(535)	3.4	(430)	4.1

technologists, and social workers lead the list of occupations with the largest numbers of assaults. Overall, health-related occupations accounted for approximately 44% of all assault-related claims over the entire period. Declines in claims numbers were seen among every health care related occupation over this period.

Since working alone has been identified as a risk factor for workplace violence in the service industries, it is worth looking at data on staffing levels for the highest risk industries. These are shown in Table 6. One notes that the number of FTE workers fell over the study period for psychiatric hospitals and skilled nursing care facilities, the two industries with the highest assault rates among all the health care industries. Since the number of these facilities did not decline, staffing levels in existing institutions fell.

DISCUSSION

Across the years of this study, the workers' compensation data consistently rank social services as the highest risk major industry, in terms of claims rate, followed by health services. These data also show health services as accounting for the majority of all work-related assault injuries. Roughly half of all assaults resulting in claims for workers' compensation happened to workers in occupations related to health care and social services. This pattern is consistent with that reported in previous studies (Bensley et al., 1997; NIOSH, 1996). In reviewing the injury descriptions in the claims database, it is apparent that assaults in the health care and social service industries are primarily encounters between caregivers and patients or between social

TABLE 5
Occupations with the Largest Number of Workers' Compensation Claims Related to Assaults and Violence Claims, Washington State, 1992–1995

Code	Occupation	1992 No.	1993 No.	1994 No.	1995 No.	Average 1992–1995 No.
447	Nursing aides/orderlies	463	564	474	374	469
446	Health aides, excluding nurses	392	332	334	232	323
208	Health technologies, n.e.c.	130	140	118	110	125
174	Social workers	100	116	91	103	103
95	Registered nurses	106	95	106	83	98
207	Licensed practical nurses	105	80	101	87	93
426	Guards/police, excluding public service	98	77	91	93	90
418	Public service, police/detectives	110	78	82	86	89
260	Cashiers	65	58	61	43	57
424	Corrections officers	50	49	49	41	47
19	Managers and administrators, n.e.c.	42	47	43	45	44
469	Personal service occupations	22	43	38	25	32
468	Child care workers, excluding private households	22	25	38	37	31
	Total	2647	2635	2511	2320	2529

Note: Numbers calculated using both time-loss and medical-only cases. Assault- and violence-related claims were defined as those with Z16.2 type of event codes 023, 025, 026, 027, and 502; and any source codes except 0200, 0201, 0230, 0240, 0250, 0270, and 5910.

service providers and clients. As earlier studies suggest, important factors putting these workers at risk include inadequate staffing levels and training; working in isolation; and working at the client's place of residence, which is common in social services (Bensley et al., 1997; Barab, 1996; Simonowitz, 1996). Policies of deinstitutionalization of patients, which in turn place social service workers at greater risk, are also faulted for increasing workplace risks for health care workers by increasing the proportion of the institutionalized population with more serious diagnoses.

In two respects this report may not fully capture the true scale of the problem of work-related assault and violence. First, the workers' compensation data source underestimates the number of injuries related to assaults and violence. The workers' compensation data only include incidents from among the self-insured employers that result in at least four lost workdays. Second, the workers' compensation data do not include self-employed workers or those covered under federal workers' compensation programs. A workplace assault only develops into a workers' compensation claim if it results in an injury at least requiring medical treatment. In addition, these data are subject to the various disincentives for employees and employers to report workplace injuries. The scale of this underreporting can be appreciated by comparing the estimates of assault frequencies from different sources.

TABLE 6
Full-Time Employment in Health Care and Social Service Industries with the Highest Risk of Assault-Related Injury, State Fund Employees, Washington State, 1992–1995

Industry	Full-Time Equivalent Workers				
	1992	1993	1994	1995	Average 1992–1995
1. Psychiatric hospitals (SIC 8063)	2,958	2,878	2,677	2,614	2,782
2. Residential care (SIC 8361)	7,280	7,256	7,025	6,897	7,115
3. Skilled nursing care (SIC 8051)	17,338	17,406	15,814	15,365	16,481
4. Nursing/personal care (SIC 8059)	1,612	1,893	1,898	2,016	1,855
5. Job training services (SIC 8331)	3,890	3,853	3,943	4,239	3,983
6. Administration of social and manpower programs (SIC 9441)	4,925	5,582	6,142	6,300	5,737
7. General medical/surgical hospitals (SIC 8062)	1,979	1,934	1,784	2,242	1,985
8. Child day care services (SIC 8351)	7,819	8,364	8,788	9,159	8,533
9. Specialty outpatient clinics (SIC 8093)	3,115	3,317	3,563	3,413	3,352
10. Home health care services (SIC 8082)	3,099	3,390	3,563	3,963	3,504

For example, in a study of workplace violence at two Washington psychiatric facilities, the researchers found in their survey of staff that the ratio of assaults reported in the survey to that found in the "incident reports" filed by the hospitals was 5:1. They also found that the number of workers' compensation claims filed by staff was less than 5% of all assaults reported in the survey (Bensley et al., 1997). In addition, the lack of a formal coding procedure at L&I for assault- or violence-related claims might lead to an as yet unknown number of assault cases miscoded in such a way as to be missed by our criteria. Finally, data from the National Crime Victimization Survey of 1987–1992, published by the Department of Justice, estimate that there were over 1 million incidents of violent assault occurring annually while the victims were at work (Bureau of Justice Statistics, 1994; Bachman, 1996). This is compared to the annual average of roughly 22,000 lost workday assault-related cases reported by the BLS survey of occupational injuries and illnesses.

In conclusion, assault-related injury in Washington State continues to be a significant contributor to workplace morbidity in health care settings. This analysis suggests, however, that the circumstances in which these incidents occur are predictable, and therefore that appropriate intervention strategies can succeed in preventing many of these incidents.

ACKNOWLEDGMENT

The authors would like to thank Lisann Rolle, Washington State Department of L&I, who prepared and provided data from the 1992–1995 U.S. Department of Labor,

BLS Statistics Census of Fatal Occupational Injuries and Survey of Occupational Injuries and Illnesses. Emily Allen, Susan Sama, and Martin Cohen provided helpful comments on an earlier draft of this document.

REFERENCES

Alexander B, Franklin G, Wolf M. The sexual assault of women at work in Washington State, 1980 to 1989. *Am J Public Health* 1994 84:640–642.

Bachman R. Epidemiology of violence and theft in the workplace. *Occup Med* 1996 11:237–241.

Barab J. Public employees as a group at risk for violence. *Occup Med* 1996 11:257–267.

Bensley L, Nelson N, Kaufman J, Silverstein B, Kalat J, Shields J. Injuries due to assaults on psychiatric hospital employees in Washington State. *Am J Ind Med* 1997 31:92–99.

Bureau of Labor Statistics (BLS). U.S. Department of Labor, Survey of occupational injuries and illnesses, 1992–1995. Washington, D.C., 1993–1996.

Castillo D, Jenkins E. Industries and occupations at high risk for work-related homicide. *J Occup Med* 1994 36:125–132.

Erickson R. Retail employees as a group at risk for violence. *Occup Med* 1996 11:269–276.

Kraus J, McArthur D. Epidemiology of violent injury in the workplace. *Occup Med* 1996 11:210–217.

Nelson N. Violence in Washington workplaces, 1992. Technical Report 39-1-1995. Safety and Health Assessment and Research for Prevention, Washington State Department of Labor and Industries, Olympia, WA, May 1995.

National Institute for Occupational Safety and Health (NIOSH). Health workers at risk for workplace violence. *Public Health Rep* 1996 111:477–485.

Simonowitz J. Health care workers and workplace violence. *Occup Med* 1996 11:277–291.

Washington State Department of Labor and Industries (L&I). Occupational injury and illness survey, 1992–1995. Olympia, WA, 1996b.

Toscano, G. Workplace violence: an analysis of Bureau of Labor Statistics data. *Occup Med* 1996 11:227–235.

Bureau of Justice Statistics, U.S. Department of Justice. Criminal victimization in the United States, 1992. 1994. National Criminal Justice publication 145125.

APPENDIX

For the purposes of this analysis, assault- and violence-related claims were defined as those with type of event and exposure codes 023, 025, 026, 027, or 502 (respectively, kicked by, bit by, struck, struck by another person in the act of a crime, or shot), and any source code *except* 0200, 0201, 0230, 0240, 0250, or 0270 (which are cases of assaults by animals) *or* 5910 (injury caused by victim). In addition to these, claims that were coded as type 028 (stabbed) were defined as assaults if and only if their source was coded as 5900 (person other than injured). This was done so as not to erroneously include many self-inflicted, accidental stabbings. Finally, claims recorded as type 029 (struck by, not elsewhere classified) and source 5900 were found to comprise a mixture of assaults and "accidental collisions" between the claimant and another person. It is possible that coding practices differ across coders or may change over time due to changes in training practice. A sample of

Codes Selected to Represent Assault- and Violence-Related Incidents

Code	Definition	Criteria
	Source Number	
0200, 0201, 0230, 0240, 0250, 0270	Animals of various kinds	Excluded
5900	Person, other than injured	Included
5910	Person, injured	Excluded
5999	Person, unspecified	Included
6000	Firearm	Included
All other codes	All other sources	Included
	Type Number	
023	Kicked by	Included
025	Bit by	Included
026	Struck or beaten by fellow worker, patient, etc.	Included
027	Struck or beaten by person in the act of a crime	Included
028	Stabbed by	Included only when source is 5900
029	Struck by, not elsewhere classified	Assault proportion estimated (see text)
502	Shot by another person	Included

these cases revealed that approximately 63% of these events were nonassaults. This proportion was used to estimate the total number of cases coded as type 029 and source 5900 that were assaults, and this was added to the total number of all cases. No attempt was made, however, to assign these cases to industries. Thus the tables that present industry and occupation figures exclude all cases coded as type 029.

9 Caring 'Til It Hurts: How Nursing Home Work Is Becoming the Most Dangerous Job in America*

CONTENTS

* This study was prepared by the Service Employees International Union (SEIU) and is reprinted with permission. SEIU is the largest union of health care workers in the United States, with 475,000 health care members working in nursing homes, hospitals, health maintenance organizations (HMOs), home care agencies, and other facilities. With a total of 1.1 million members in the United States and Canada, SEIU is the third largest and fastest growing union in the American Federation of Labor and Congress of Industrial Organization (AFL-CIO).

EXECUTIVE SUMMARY

What is the most dangerous job in America? Mining? Construction? Trucking? Working in a steel mill? No. More dangerous than all these—and fast becoming the most dangerous job in the United States—is nursing home work.

Ironically, nursing home work is also one of the fastest growing jobs in America. While working conditions in many other industries have gradually improved over the last decade, nursing homes have become far more dangerous places to work. Caregivers suffer an epidemic of crippling workplace injuries.

These injuries, devastating as they are to a growing workforce of committed caregivers, also point to a broader and equally dangerous problem. That problem is a developing crisis in staffing and conditions in nursing homes nationwide that threatens the quality of care for millions of America's most vulnerable citizens. It is a crisis that is occurring just as the $85 billion, mostly taxpayer-financed industry is experiencing its most radical transformation in a quarter century.

The Service Employees International Union (SEIU) has found that:

- Occupational illness and injury rates for nursing home workers are higher than for workers in other industries with well-documented hazards, such as mining and construction.
- More than 18% of all nursing home workers are injured or become ill on the job each year—more than twice the general rate of private sector workers.
- Occupational illness and injury rates for nursing home workers increased by 57% between 1984 and 1995, with more than 200,000 injuries reported in the industry every year.
- Of those injured nursing home workers who must take time off, more than a quarter require more than two work weeks to recover. Less than a third are able to return within 1 or 2 days.
- Of the 20 fastest growing industries in the United States, nursing homes have the highest rate of occupational illness and injury.

BACK INJURIES ARE EPIDEMIC

Back injuries, widely agreed to be among the most serious and costly of workplace injuries, are most commonly suffered by nursing home workers.

- While back injuries account for 27% of all injuries reported in the private sector, they account for 42% of all those in nursing homes.
- Nurses' aides, who provide most of the care in nursing homes, are particularly at risk. Injuries to the back and trunk account for more than half of all injuries to nurse aides working in nursing homes.

UNDERSTAFFING CAUSES INJURIES TO WORKERS AND DIMINISHES QUALITY OF CARE

The shift to prospective payment in both private and public health insurance has led to earlier hospital discharges and an overall increase in the acuity levels—the "sickness levels"—of nursing home patients. Unfortunately, staffing levels have not increased to match the increased workload.

- Data from the federal On-Line Survey Certification and Reporting System (OSCARS) show that the number of nurses' aide hours per resident day only increased from 2.0 in 1992 to 2.1 in 1995, an insignificant amount.
- Understaffing, according to a 1996 National Academy of Sciences' prestigious Institute of Medicine (IOM) report, leads to injuries, which leads to further understaffing. Nurses' aides are often forced to lift residents alone when assistance is not immediately available. With sicker and more dependent residents, nursing homes have become more hazardous in terms of injuries.
- The IOM report also states that research has demonstrated "a positive relationship between nursing staff levels and the quality of nursing home care, indicating a strong need to increase the overall level of nursing staff in nursing homes."

EMPLOYERS DO NOT TAKE SAFETY CONCERNS SERIOUSLY

Rather than hiring additional staff or redesigning work to eliminate hazards, nursing home employers have tended to favor gimmicks that do little to prevent injuries. Much of the focus has been on teaching proper lifting techniques, which are of limited use in patient care settings. In addition, the SEIU has also found that:

- Inspections of nursing homes by the Occupational Safety and Health Administration (OSHA) have often found that safety training programs for nurses' aides are not effective. Nurses' aides generally do not use the techniques outlined in the program, either because there are not enough staff members on hand or because they have not received adequate instruction in the techniques.
- Back belts are a favorite device of nursing home operators. However, there is no evidence that they reduce the hazards of repeated lifting, pushing, twisting, and bending.
- Many employers sponsor programs such as "safety bingo" that encourage workers not to report injuries in exchange for the chance to win a television set or other merchandise.

UNSAFE WORKING CONDITIONS EXACT A HIGH COST FROM TAXPAYERS

Because public programs pay for most nursing home care, taxpayers ultimately bear the burden of unsafe working conditions in the nursing home industry. Taxpayers pay for more than half of the nursing home care provided in the United States, primarily through the Medicaid program.

Nursing home industry executives often resist hiring additional staff or purchasing safety equipment because of the cost. Nevertheless, the truth is that running an unsafe workplace increases costs in a variety of hidden—as well as more obvious—ways.

- The staggering workers' compensation insurance rates paid by nursing home employers are a reflection of the magnitude and severity of injuries and illnesses in nursing homes. The nursing home industry as a whole paid close to $1 billion in workers' compensation insurance payments in 1994.
- The high rate of injury in nursing homes has contributed to high rates of turnover—as high as 100% per year at some facilities. Constant turnover forces employers to invest resources in recruitment and training of new workers.
- Conditions that contribute to worker injuries, such as understaffing, also add to patient injuries. The nursing home industry spends $1.2 billion to heal preventable decubitus ulcers (bedsores) caused by lack of nutritional hydration, mobility, and cleanliness. The industry spends $4.3 billion on incontinent care because residents are not toileted frequently enough.

A number of academic studies have shown that there are cost-effective interventions that can reduce the number of injuries in nursing homes and save money for employers, patients, and taxpayers.

RECOMMENDATIONS

Based on the information presented in this study, the SEIU makes the following recommendations:

Expand and improve the OSHA special emphasis program for nursing facilities—In August 1996, OSHA announced a new "special emphasis program" aimed at nursing facilities. This seven-state initiative involves educational outreach to employers and workers in the industry, teaching them how to identify and abate various safety and health risks. OSHA needs to expand the program to more states and to be much more aggressive in taking enforcement action at facilities with the largest number of documented safety and health problems.

Establish an ergonomics standard—OSHA is in the process of developing an ergonomics standard for private sector employers. Such a standard would require a work site risk analysis and the evaluation and implementation of feasible methods of preventing and reducing those risks. It should also include a program of prompt treatment for workers who are injured—even those who report just the first signs of a possible injury. Education and training in the prevention of injuries should also be included. OSHA is moving forward on the development of a proposed ergonomic standard. Strong opposition from industry continues. OSHA should take a firm stand for workplace safety and ensure that the proposed standard covers all industries where workers continue to suffer from crippling strain and sprain injuries.

Improve staffing standards—The cornerstone of worker safety and quality resident care is adequate staffing. Existing federal staffing standards, enacted almost a decade ago, are inadequate and do not regulate staffing levels for nurses' aides,

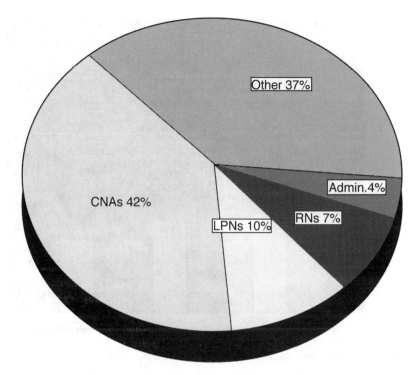

FIGURE 1   Nurses' aides are the largest part of the nursing home workforce.

who provide the bulk of direct resident care. The SEIU recommends the establishment of acuity-based staffing ratios, with specific minimum ratios for nurses' aides.
The U.S. Health Care Financing Administration (HCFA) should monitor the relationship of staffing to acuity levels, resident care, and worker and patient injury rates
on an ongoing basis. As a first step, the National Institute for Occupational Safety
and Health (NIOSH) should conduct a comprehensive study of the relationship
between short-staffing and workplace injuries in nursing facilities.

INTRODUCTION

The nursing home industry employs roughly 1.3 million workers who provide round-
the-clock care and assistance in the activities of daily living to chronically ill and
disabled individuals. The nursing home workforce is disproportionately female
(87%), minority (30%), and low-wage earners. The average nursing home worker
earns just over $15,000 a year, and many work under stressful and hazardous
conditions that put their own health, as well as the health of their residents, at risk.*

   Unlike hospitals, where registered nurses (RNs) provide most of the hands-on
patient care, certified nurses' aides (CNAs) provide most of the care in nursing
homes (see Figure 1). CNAs constitute 42% of the nursing home workforce.

---

* All figures are from Bureau of Labor Statistics (BLS), *Employment and Earnings*, monthly.

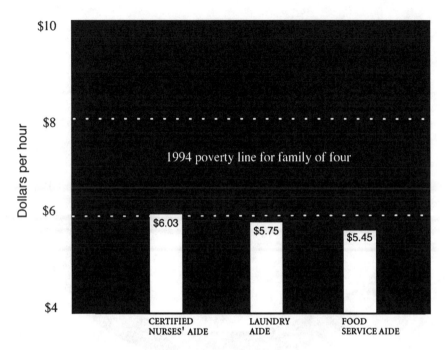

FIGURE 2　Wages for most nursing home jobs are below the poverty line.

Many of the workers employed in nursing homes earn poverty-level wages. As Figure 2 makes clear, wages for nurses' aides and a range of other nursing home job classifications are below the poverty line for a family of four.

Nursing home workers not only suffer the insult of below-poverty-level wages, but also they also must work in what is rapidly becoming one of the most dangerous jobs in America. While most people believe that nursing homes are safe and clean, they are health care facilities, after all; as stated earlier, working in a nursing home is actually more dangerous than working in a coal mine, a steel mill, a warehouse, or a paper mill.

While working conditions in many other industries have gradually improved over the past decade, those in nursing homes have become more dangerous. Between 1984 and 1995, the injury rate for nursing home workers increased by 57%. Today, nursing home workers generally are injured at more than twice the rate of private sector workers.

The nursing home industry also holds the distinction of being the most dangerous growth industry in the United States. Of the 20 fastest growing industries in the United States, nursing homes have the highest illness and injury rate. The nursing home industry is expected to add more than 750,000 new jobs by the year 2005.

While back injuries—widely agreed to be among the most serious and costly of injuries—account for a quarter of all injuries reported in the private sector, they account for almost half the injuries in nursing homes.

The consequences of this epidemic of workplace injuries extend to all Americans: to workers, who suffer debilitating and disabling injuries; to their families, whose standard of living may be severely reduced by a breadwinner's disability; to nursing home employers, who pay ever-increasing costs for workers' compensation, for retraining and recruitment, and for decreased productivity; and to taxpayers, whose tax dollars finance Medicaid and Medicare, which pay for most of the nursing home care in the United States.

For too long, serious back injuries and strains have been viewed as an inevitable part of nursing home work—"just part of the job." We can no longer afford to take that view. It is time for the nursing home industry, federal and state legislatures, government agencies, and nursing home workers to join together to defeat the terrible epidemic of crippling workplace injuries in nursing homes.

If we fail to act, we risk losing the committed, skilled workers who provide long-term care to millions of Americans. These workers know that the longer they continue to work in a nursing home, the more likely they will suffer a permanently disabling injury. Not only will the industry be unable to retain experienced workers, but also it will become increasingly difficult to recruit new staff members.

This SEIU report reviews the extent and the character of injuries in nursing homes, describes their major causes, and illustrates their effects on workers and patients. It also provides estimates of the cost that this new epidemic of workplace injuries imposes on the public.

DANGERS OF NURSING HOME WORK

In 1994, the injury and illness incidence rates for nursing homes was an alarming 18.2 per 100 full-time workers—greater than those for coal mining (6.2), blast furnaces and steel mills (11.9), warehousing and trucking (13.8), and paper mills (7.5).*

Nursing homes have become much more dangerous over the past decade. Between 1984 and 1995, the illness and injury rate increased from 11.6–18.2 per 100 full-time workers—a 57% increase. The bitter irony, as Figure 3 makes clear, is that injury rates for nursing homes have been rising while rates for other high-risk industries like manufacturing and mining have been falling.*

These injuries are not minor. Data from the Bureau of Labor Statistics (BLS) shows that 50% of injured nursing home workers must take days off work or work "light duty" to recover. This is a higher rate than that for the private sector as a whole (47%) or for the manufacturing (45%) or construction (47%) industries. Of those nursing home workers who must take days off work, more than a quarter require more than 2 weeks (11 or more working days) to recover (see Figure 4). Less than a third are able to return within 1 or 2 days.**

In 1995, nursing homes reported 246,900 injuries on the job. The nursing home industry ranked third in the total number of injuries, exceeded only by eating and drinking establishments and hospitals.[1]

---

* Data are from Bureau of Labor Statistics web site at http://www.bls.gov/. In most cases, incidence rates are the number of illnesses or injuries (or both) per 100 full-time equivalent (FTE) workers.
** Data are from Bureau of Labor Statistics web site at http://www.bls.gov/. In most cases, incidence rates are the number of illnesses or injuries (or both) per 100 full-time equivalent (FTE) workers.

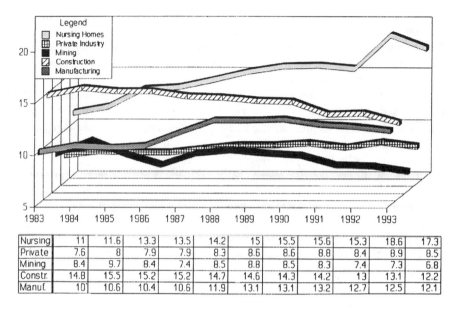

	1983	1984	1985	1986	1987	1988	1989	1990	1991	1992	1993
Nursing	11	11.6	13.3	13.5	14.2	15	15.5	15.6	15.3	18.6	17.3
Private	7.6	8	7.9	7.9	8.3	8.6	8.6	8.8	8.4	8.9	8.5
Mining	8.4	9.7	8.4	7.4	8.5	8.8	8.5	8.3	7.4	7.3	6.8
Constr.	14.8	15.5	15.2	15.2	14.7	14.6	14.3	14.2	13	13.1	12.2
Manuf.	10	10.6	10.4	10.6	11.9	13.1	13.1	13.2	12.7	12.5	12.1

FIGURE 3   Nursing home injury rate soars.

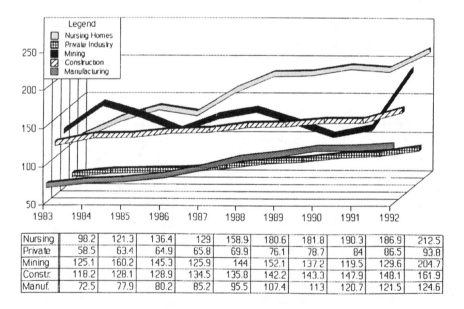

	1983	1984	1985	1986	1987	1988	1989	1990	1991	1992
Nursing	98.2	121.3	136.4	129	158.9	180.6	181.8	190.3	186.9	212.5
Private	58.5	63.4	64.9	65.8	69.9	76.1	78.7	84	86.5	93.8
Mining	125.1	160.2	145.3	125.9	144	152.1	137.2	119.5	129.6	204.7
Constr.	118.2	128.1	128.9	134.5	135.8	142.2	143.3	147.9	148.1	161.9
Manuf.	72.5	77.9	80.2	85.2	95.5	107.4	113	120.7	121.5	124.6

FIGURE 4   Lost workdays on the rise.

Another dubious distinction for the nursing home industry, as mentioned earlier, is that it is the most dangerous growth industry in the United States. Of the 20 fastest growing industries in the United States, nursing homes rank 11th if ranked by rate of growth. However, if those 20 industries are ranked by injury incidence rates,

TABLE 1
Nursing Homes Are the Most Dangerous Fast-Growing Industry

Industry	Annual Rate of Job Growth 1994–2005 (%)	Projected Jobs Added 1994–2005 (000s)	Injury and Illness Incidence Rate 1995
Nursing and personal care facilities	3.5	751	18.2
Miscellaneous transport services	3.0	75	16.3
Residential care	5.7	498	11.7
Job training and related	3.4	127	10.6
Water and sanitation	3.2	87	10.5

Note: Industries among the 20 fastest growing, ranked by incidence rates.

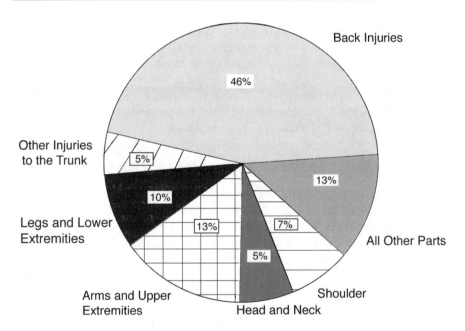

FIGURE 5   Back injuries are most common for nurses' aides. (*Source:* BLS, 1992.)

nursing homes rank first. Table 1 shows the five most dangerous fast-growing industries.[2]

BACK INJURIES PREDOMINATE

Nursing home workers are more likely to be injured on the job than other workers, and their injuries more often involve the back, widely agreed to be among the most serious and costly of injuries. Data from the 1994 BLS survey of occupational injuries found that back injuries make up 27% of injuries reported in all private industry, but account for 42% of all injuries in nursing homes (see Figure 5).[2]

The risk of back injury is slightly higher if only nurses' aides, who provide the bulk of patient care, are considered. The BLS found that back injuries account for 46% of all injuries to nurses' aides in nursing homes. Taken together, back and shoulder injuries account for 53%. Most of these injuries are classified as "sprains and strains"—82% of the reported injuries fell into this category. [2]

Reporting patterns may also lead to underestimating the true risk. Strains are usually reported as work related only if they are acute. In nursing homes, stressful activities that do not necessarily cause acute injury lead to cumulative trauma that can end in debilitating injury or disease. Unless they can point to a single causal incident, workers are discouraged from reporting the injury. In many facilities, employer safety programs such as "safety bingo" (see pp. 2 and 10) give workers incentives not to report their injuries.

Nursing home workers are also at greater risk of workplace violence than workers in other industries. Of the 20,438 workers injured in 1994 in nonfatal assaults that required days away from work, 4,905 (24%) were employed in nursing homes—a higher rate than any other industry. In the overwhelming majority of cases, the assailant was a resident of the facility where the worker was employed. The attacker was usually suffering from some form of dementia. The median days away from work was 5 days. Slightly more than a fifth of the nonfatal assaults resulted in injuries severe enough to require 21 days or more away from work.[3] Understaffing puts workers at greater risk of assault because they are alone with residents, with no one to back them up if a resident becomes violent.

WHY IS NURSING HOME WORK SO HAZARDOUS?

As far back as 1980, the injury and illness incidence rate for nursing homes was 10.7%—higher than the rate of 8.7% for all private industry. Much of the reason for this is the physically demanding nature of nursing home work. Nurses' aides lift and turn patients, help them in and out of baths, make beds, and take residents to and from the toilet. Housekeepers vacuum, dust, mop, clean walls and windows, and collect trash. Dietary workers carry heavy boxes and pans, clean their work areas, and cut and prepare large amounts of food. Laundry workers lift wet, heavy linens. They load and unload large washing machines and dryers. All nursing home workers use large and small muscles over and over again, often to move bodies or objects that weigh more than they do.

The more a worker must handle patients, the greater the risk of injury. OSHA found that 81% of back and shoulder injuries to nurses' aides in nursing homes were caused by residents. Of all back and shoulder injuries, 47% were caused by "over-exertion in lifting" and 17% were caused by "overexertion in holding, carrying, or turning," all occurring when handling a patient.[2]

Nurses' aides are particularly at risk. OSHA inspections in 1992 of a number of nursing home chain facilities in Pennsylvania found that short-staffing was forcing nurses' aides to perform many patient lifts and transfers alone. On many shifts, a single nurses' aide would perform up to 40 lifts and transfers of residents weighing as much as 260 pounds—all without the help of another employee or mechanical hoist. In short, many nursing aides were lifting over *10,000 pounds per shift*.[4]

These amounts far exceed recommended safety levels. The International Labor Organization and NIOSH recommend that manual lifting be limited to about 50 pounds twice an hour for objects with a center of gravity about 15 inches from the lifter.[5] (The center of gravity is at least 15 inches or more away from nurses' aides when they are lifting residents.)

RISKS TO NURSING HOME WORKERS HAVE BEEN INCREASING

While nursing home work has always been hazardous, there is evidence that conditions for nursing home workers have become much more perilous. A number of factors have converged to make nursing home work much more dangerous to workers than it was 10 or 15 years ago.

The most important factor has been a reduction in the length of hospital stays, especially for the Medicare population, which many analysts trace to the federal government's introduction of prospective payment into Medicare in 1983. Other factors, such as improvements in medical technology, have also played a role. The average hospital stay for an elderly patient in 1993 was 1.9 days shorter than in 1983, a 19% reduction. The average stay for the adult population as a whole fell by 10% over that period.[6]

With patients leaving the hospital earlier, the burden of providing "subacute" and specialty care is increasingly on nursing home workers. According to the American Health Care Association (the nursing home industry's trade association), there are currently between 10,000 and 15,000 subacute beds and more than 60,000 beds in specialty care units nationally. Problems of definition make it difficult to estimate the true size of the subacute market, but a study by Lewin-VHI put it between $600 million and $3.4 billion and between 1.2 million and 8.1 million patient days.[7]

Clinically, these changes have meant that residents tend to need more assistance with activities of daily living. An increasing number of residents also need rehabilitative services, oncology, wound care, and infusion services. Nursing home workers are also having to treat patients with more complex medical conditions such as Alzheimer's disease and acquired immunodeficiency syndrome (AIDS).

UNDERSTAFFING PUTS PATIENTS AND WORKERS AT RISK

Nursing home operators have failed to address either the problem of unsafe working conditions or the dramatic worsening of those conditions over the past few years. Staffing levels, for example, have not kept up with the changes in patient mix.

In 1996, the National Academy of Science's IOM released its long-awaited report on the adequacy of nurse staffing in hospitals and nursing homes.[8] The IOM report reviewed a number of studies of the relationship between staffing and quality of care in nursing homes. It concluded that "the preponderance of evidence from a number of studies using different types of quality measures has shown a positive relationship between nursing staff levels and quality of nursing home care, indicating a strong need to increase the overall level of nursing staff in nursing homes."

TABLE 2
Hours per Resident Day for
All Certified Nursing Facilities, 1992–1995[a]

	1992	1993	1994	1995
RNs	0.4	0.4	0.5	0.5
Licensed practical nurses	0.6	0.7	0.7	0.7
Nurses' aides	2.0	2.1	2.1	2.1
Total nurse hours[a]	3.0	3.1	3.1	3.3

[a] Totals may not add due to rounding.

Source: Harrington et al., 1996.

The IOM report also looked at the relationship between staffing and health and safety problems. In a relatively strong statement, it concluded that back injuries are related to inadequate staffing:

Staffing levels in nursing homes have not kept pace with the increased demand for more and better trained personnel. The case mix of nursing home patients is increasing in complexity as hospitals discharge patients early and transfer them to nursing homes. With sicker and more dependent patients than in the past, nursing homes have become more stressful and hazardous in terms of injuries. This situation is reflected in the high turnover among NAs [nurses' aides] who do most of the heavy lifting. Understaffing (both quantitative and qualitative) leads to injuries, which leads to further understaffing and the needs of the patients go unmet. Often NAs are forced to lift residents alone when assistance is not immediately available (Reference 8, p. 174).

The report also concluded that more and better training was necessary:

The situation clearly suggests the importance of and need for more aggressive training related to the use of lifting devices and lifting teams especially for new employees, including ergonomic training in lifting techniques to prevent back injuries .... All personnel giving direct care ... should have annual training in lifting and transferring patients. Such efforts would improve the quality of life for health care workers and would represent a significant savings to the health care industry (Reference 8, pp. 174–175).

While the link between staffing and quality has become increasingly clear, nursing home operators have made only minimal moves to increase staffing. Table 2 summarizes a 1996 study[9] using data from the federal OSCARS that tracks nursing home staffing. As this table makes clear, nursing homes have increased the number of RNs on duty, but they have not significantly increased the number of nurses' aides.

While increasing RN staffing will probably have a positive impact on mortality rates of nursing home residents, it will do little to reduce the number of resident and worker injuries that occur during the lifting and transferring of residents. This work is generally performed by nurses' aides.

When working in a chronically understaffed facility, the pressure to pull one's own weight is enormous, even when workers are not physically up to the task because of injury or fatigue. Workers often are forced to lift and turn patients by themselves or make difficult choices between remaining with patients who need immediate assistance or leaving to help their co-workers. Many workers injure themselves further by doing lifts to avoid overburdening injured co-workers. More experienced workers can be injured trying to help new employees.

EMPLOYERS FAVOR GIMMICKS RATHER THAN REAL SOLUTIONS

Rather than hiring additional staff or redesigning work to eliminate hazards, nursing home employers have tended to favor gimmicks that do little to prevent injuries. In 1987, for example, one nursing home chain distributed a training program to all its facilities entitled Lift with Care. The program was presumably designed to teach "proper lifting techniques" to nursing home staff. While the concept of good body mechanics is the battle cry of trainers, such techniques—back straight, legs bent, and objects between legs or close to body—are not easily applied to nursing home work.

For example, a single lift may require leaning over a bed to lift a resident and then holding the individual while turning 90 degrees and lowering him or her into a chair. Patients may have to be lifted over chair arms, from a low chair to a higher bed, or even up from the floor. Transfers often involve contortion of the trunk, uneven weight distribution, and exertion from awkward postures. The additional stress of sustaining awkward positions—as when stooping to feed a resident or pulling a wheelchair while at the same time moving an individual—further stretches and weakens muscles.

OSHA inspections of nursing homes have revealed that safety training programs for nurses' aides are not effective. Although training programs like Lift with Care have several components, in many cases only the 25-min video is used. Nurses' aides seldom use the techniques outlined in the program, either because the lifts require more staff than are on hand or because instruction on the techniques is inadequate.*

Another innovation is to require workers to wear girdle-like devices called "back support systems" or "back belts" designed to keep workers' backs rigid. Although they have not been comprehensively tested in nursing homes, there is no evidence to date that such devices are effective in preventing injuries. A 2-year study by the Back Belt Working Group of NIOSH concluded that "back belts do not mitigate the hazards to workers posed by repeated lifting, pushing, pulling, twisting, or bending." The working group "does not recommend the use of back belts to prevent injuries among uninjured workers, and does not consider back belts to be personal protective equipment."[10]

More significant than what nursing homes do is what they do not do. Nursing aides report that the nursing care plans for each patient usually fail to provide written

---

* These examples are taken from inspection reports from several Pennsylvania nursing homes submitted by Dr. Bernice Owen to the Occupational Safety and Health Administration in 1991.

directions about the number of staff needed to transfer the patient or the safest way to do so, and rarely do supervisors provide this information.*

Nursing home operators have also resisted investing in mechanical lifting equipment that could reduce the hazard to nursing home workers. Although advanced equipment is available, most nursing homes continue to use outmoded lifting equipment. In many cases, the lifts are used for weighing residents and are unavailable for use in transfers. Many nurses' aides also report that they seldom use lifting equipment, even when it is available, because it takes too much time, it is not safe, or they do not feel comfortable using it.* These problems have been confirmed by a number of academic studies.[11]

The bottom line is that the nursing home industry has not committed itself to a comprehensive program that would prevent workplace injuries. Many employers continue to deny the extent of the problem, contesting OSHA citations and refusing to negotiate legitimate solutions. External pressure from organized labor and government agencies will undoubtedly be needed to force the industry to make significant changes in the way it treats workplace hazards and injuries.

One of the most egregious employer initiatives is so-called "safety bingo."

"The company constantly claims their compensation insurance is too high. But instead of putting on more staff, they introduced 'safety bingo.' The grand prize was a 19-inch color remote-controlled television set. They drew a bingo number whenever a week passed without a reported injury. If an injury was reported, everyone would have to throw their bingo card away and start all over. Some people were so committed to winning the television that they wouldn't report injuries."[12]

SAFETY SAVES MONEY

Employers in the nursing home industry often resist hiring additional staff or purchasing safety equipment because of the cost. However, running an unsafe workplace does not save money in the long run. It actually increases costs in a variety of hidden—as well as more obvious—ways.

Workers' compensation—The staggering workers' compensation insurance rates paid by nursing home employers are a reflection of the magnitude and severity of injuries and illnesses in nursing homes. In a number of states, the insurance rate for nursing homes is twice or even three times the rate for an average private sector employer. The SEIU has estimated that the nursing home industry as a whole paid close to $1 billion in workers' compensation insurance costs in 1994.**

Injuries drive workers from the profession—Turnover in the nursing home industry is very high—in excess of 100% at some facilities.[8] New workers are shocked and dismayed at the frantic pace they are expected to maintain. Experienced

---

* These examples are taken from inspection reports from several Pennsylvania nursing homes submitted by Dr. Bernice Owen to the Occupational Safety and Health Administration in 1991.
** This was estimated as 6% of total payroll for the entire nursing home industry. Total payroll was estimated as the average weekly wage for the industry, multiplied by 52 (weeks), multiplied by the total number of workers in the industry (1.7 million). All figures except for 6% (an average based on available state data) is from the Bureau of Labor Statistics.

workers burn out when they are unable to provide the care their residents need. Turnover forces employers to invest in recruitment and training of new workers.

Reduced tax base—People on workers' compensation do not pay income taxes. They have less money available to spend on taxable goods. They may require vocational retraining through state-funded programs as well as other forms of assistance.

Health insurance redlining shifts costs to other industries and workers— Because nursing homes are such hazardous places to work, many insurance companies refuse to sell health insurance to nursing home operators. A University of Michigan survey of insurance companies found that two thirds of those surveyed listed nursing homes among a group of employers ineligible for coverage under any circumstances, and an additional quarter of those surveyed would only offer coverage with certain restrictions.[13] When nursing home workers are unable to obtain health insurance at work, the cost of their medical care is shifted to other employers and workers.

Injuries to patients drive up the cost of care—Workers are not the only ones put at risk by unsafe working conditions. Patients also suffer. As stated earlier, the industry spends $1.2 billion to heal preventable decubitus ulcers (bedsores) caused by lack of nutritional hydration, mobility, and cleanliness; and $4.3 billion on incontinent care because residents are not toileted frequently enough.[14]

TAXPAYERS ULTIMATELY PAY THE COST OF UNSAFE WORKING CONDITIONS

Again, because public programs pay for most nursing home care, taxpayers ultimately bear the burden of unsafe working conditions in the nursing home industry. The share of nursing home costs paid for by taxpayers has risen from 51% in 1990 to 55% in 1994. The Medicaid program is the single largest source of money for nursing homes, accounting for more than 74% of nursing home revenue (see Figure 6).[15]

Of the $1 billion paid out by nursing home operators in workers' compensation insurance premiums, for example, well over half that amount was paid by taxpayers. Similarly, the industry was able to shift to taxpayers 55% of the $1.2 billion spent on healing preventable bedsores, the $2.6 billion spent on treating residents injured in falls, and the $4.3 billion spent on treating conditions associated with incontinence.[8]

COST OF SAFE WORKPLACE OVERESTIMATED

The preceding data suggest that employers have significantly underestimated the costs associated with operating an unsafe workplace. The irony is that they also overestimate the cost of having a safe workplace.

A typical nursing home with 100 workers pays between $50,000 and $100,000 a year in workers' compensation insurance payments. These funds could also be used to invest in technology that would make lifting and transferring patients less hazardous. One year's worth of workers' compensation payments for a typical nursing home would pay for 10 to 15 mechanical devices, which would allow staff to lift and transfer patients without having to lift too much weight. A wide range of affordable devices are available:

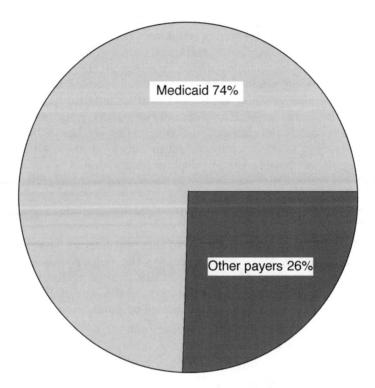

FIGURE 6   Taxpayers pay most of the cost of nursing home care.

- Chairs are available that can double as a toilet and a shower stool to eliminate unnecessary transfers. Currently, patients are often lifted from bed to wheelchair, wheelchair to toilet, toilet to wheelchair, wheelchair to shower, shower to wheelchair, and wheelchair back to bed. Instead, patients could be lifted from their bed to a shower chair (where toileting and showering are done) and from there back to bed, eliminating a number of lifts.
- Portable or ceiling-mounted patient hoists can substitute a mechanical device for a human lifter. Most homes possess only antiquated lifting equipment even though many safer and more versatile devices are now available.
- Hoist belts or lifting belts can allow workers to lift patients up with less strain on the worker's back by placing a belt around the hips of the patient.
- Ambulation belts are available that help stabilize patients while they are walking, reducing the risk of falls.
- Friction-reducing sliding boards can be placed beneath patients so they can slide instead of being lifted into a wheelchair.
- Other patient-handling devices are available, including roller boards, handles on sheets, turntables for pivot transfers, bath boards, and transfer devices for many different situations.

A number of academic studies have shown that these types of interventions have the potential to significantly reduce the number of injuries to nursing home workers. A study by Owen and Garg,[16] for example, examined the impact of a number of ergonomic interventions on injury rates in two nursing home units with a total of 140 beds and 57 nursing assistants. The program involved determining which patient care tasks had the highest risk of injury, modifying the jobs, and introducing new patient-transferring devices. In the 12-month period following the intervention, the back-injury rate was cut nearly in half. Of the remaining back injuries, more than 80% were not related to the stressful tasks identified in the study. Lost and restricted workdays also declined, dropping to zero during the last 4 months of the postintervention phase of the study.[16]

Many nursing home employers already have learned that working safely is good for their bottom line. A 203-bed facility in Camden, ME invested in two new lifts for residents who could not support their own weight and gait belts for residents who could walk with assistance. Staff received training from the manufacturers and reviewed proper transfer techniques. The facility's workers' compensation premium dropped from $750,000 a year to $184,000.[17]

At another facility in Erie, PA, nursing staff reported back injuries. The nursing department also had high employee turnover, primarily among nursing aides, who performed between 20 and 60 lifts per day. In 1992, the facility instituted a no-lift policy that requires nursing personnel to use lifting equipment for residents unable to move or walk on their own. A notice on each resident's bed informs the staff of what equipment, if any, is required to lift the patient. Initially, the facility used lifts with manual cranking, but within a year, switched to one with electrical cranks. Since implementing the program, the facility has had only one back injury due to lifting. Workers' compensation premiums dropped from $117,000 a year to $85,000.[17]

For nursing home operators to claim that these interventions are "too expensive" is absurd. While the industry continues to plead empty pockets, its balance sheets have never looked better. The nation's largest nursing home chains reported 1996 profits in the tens or even hundreds of millions of dollars: Vencor Inc. ($128 million), Beverly Enterprises ($75 million), Healthcare and Retirement Corp. ($59 million), and Horizon/CMS ($7 million).*

Another argument the industry makes against the introduction of lifting devices is that they are "dehumanizing." The view of Paul Willging, executive vice president of the American Health Care Association, the trade group for nursing home owners, is typical. "We don't want a nursing home to look like a meat-packing plant. It's not a factory."[18]

However, patients feel much more secure and comfortable in well-designed patient lifting equipment and are more insecure and nervous when workers awkwardly try to lift them without adequate staffing or equipment. A 1995 study by Owen and co-workers[19] asked patients to rate their level of comfort with a patient-handling task, with 0 being "very comfortable" and 7 being "extremely uncomfortable." The average

---

* Profit figures are obtained from company annual reports and 10-K forms filed with the Securities and Exchange Commission.

level of comfort for patients using assistive devices ranged from 0.2–1.2, compared to 2.0–5.0 for patients lifted by workers.[19]

RECOMMENDATIONS

Our nation can no longer afford the staggering costs imposed by the epidemic of back injuries and other cumulative trauma disorders among nursing home workers. The following steps must be taken if we are to reverse the disturbing trend in workplace injuries in nursing homes.

EXPAND AND IMPROVE THE OSHA SPECIAL EMPHASIS PROGRAM

In August 1996, OSHA unveiled a seven-state initiative to address the rising number of injuries in the nursing home industry. The seven states—Florida, Illinois, Massachusetts, Missouri, New York, Ohio, and Pennsylvania—were selected because they all have 500 or more nursing facilities.

The program involves educational outreach to employers and workers. OSHA offers free, 1-day comprehensive safety and health seminars to help participants identify and eliminate safety and health hazards, including back injuries from lifting residents; slips and falls; workplace violence; and risks from bloodborne pathogens, tuberculosis, and other infectious diseases. Enforcement action is taken against facilities where employers fail to act to reduce high injury rates.

While the program has served as a model for nursing home facilities in the target states, it has failed to improve the lot of the vast majority of nursing home workers. OSHA needs to expand the program and develop strong enforcement procedures, targeting work sites with the largest number of safety and health problems.

Additionally, OSHA and NIOSH can play an important leadership role by developing "best practice" guidelines for the industry. The agencies should continue to sponsor conferences and develop materials that define the components of a successful nursing home health and safety program.

MOVE FORWARD WITH OSHA'S PROPOSED ERGONOMICS STANDARD

Over the past few years, OSHA has been developing an ergonomics standard for private sector employers. The standard would provide clear guidance to nursing home employers on how to prevent disabling injuries. It would require that employers develop an injury prevention program that includes

- A work site analysis to identify risk factors that cause injuries (e.g., certain types of patient care tasks, such as lifts and transfers, that pose a particularly high risk)
- The evaluation and implementation of feasible methods to prevent and reduce those risk factors, including the use of mechanical lifting equipment, improved staffing levels, and better communication to staff of a resident's needs
- A program of early treatment and ongoing medical management for workers who are injured or who report early warning signs of injuries

- Education and training on how to identify and prevent injuries caused by lifting and repetitive tasks

Although OSHA was scheduled to issue a proposed version of the ergonomics standard in early 1995, vocal opposition from the business community and many Republicans in Congress blocked them. Strong opposition from industry continues. However, OSHA is now moving forward to develop a new version of the proposed ergonomic standard. It will take at least a year before the proposed standard is issued. In addition, it is not clear that the proposed standard would cover all industries, like the nursing home industry.

The SEIU recommends:

- OSHA should proceed with the development of a new version of the proposed ergonomic standard.
- OSHA must ensure that the proposed standard includes coverage of all industries, including the nursing home industry, where ergonomic hazards generate a large number of workplace injuries.

IMPROVE STAFFING STANDARDS

The cornerstone of worker safety and quality resident care is adequate staffing. While many states set minimum staffing requirements for nursing services, the standards are woefully inadequate. In many cases these standards do not differentiate between hands-on caregivers like nurses' aides and supervisory staff when defining minimum staffing requirements.

Standards at the federal level are even worse. In 1987, Congress established staffing standards for RNs in nursing homes, but failed to establish any standards for nurses' aides, who provide most patient care. In 1995 and 1996, there was a sustained attempt by Republicans in Congress to eliminate or at least sharply scale back the 1987 standards, but they were unable to overcome opposition by President Clinton and the proposals were abandoned.

While the 1987 standards were a dramatic step forward 10 years ago, it has become clear that further efforts are needed in this area. Employers are exploiting loopholes in the regulations. In some states, they are able to include administrative staff when calculating whether they meet the required staffing levels.

The SEIU recommends:

- Enforceable staffing ratios are to be linked to the acuity of residents. Such ratios should be enforced on a floor or unit level to prevent facilities from understaffing in particular units while still meeting a facility-wide target.
- There are to be specific minimum staffing ratios for nurses' aides. Since nurses' aides provide most patient care, it is important that specific standards be set for them, independent of case mix. An expert panel of the National Committee to Preserve Social Security and Medicare has proposed ratios of 1 to 8 (days), 1 to 10 (evenings), and 1 to 15 (nights). While the SEIU remains concerned that these ratios of workers to residents may not be sufficient, they are lower (i.e., better) than we observe in many facilities.

- As a first step, NIOSH should conduct a comprehensive study of the relationship between short-staffing and workplace injuries in nursing facilities.

REFERENCES

1. Bureau of Labor Statistics (BLS), *Workplace Injuries and Illnesses in 1994*, December 15, 1995, news release.
2. Data from Bureau of Labor Statistics web site at http://www.bls.gov/.
3. Bureau of Labor Statistics (BLS), *Characteristics of Injuries and Illnesses Resulting in Absences from Work, 1994*, May 8, 1996, news release.
4. Data from record evidence cited in the U.S. Department of Labor's Brief to OSH Review Commission (filed May 1, 1996) in Secretary of Labor v. Beverly Enterprises, Inc. (OSHRC Docket 91-3344, 92-0238, 0819, 1257, and 93-0724).
5. N.G. DeClercq and J. Lund, "NIOSH lifting formula changes scope to calculate maximum weight limits," *Occupational Health and Safety*, February 1993.
6. Prospective Payment Assessment Commission, *Medicare and the American Health Care System: Report to Congress*, June 1995; American Hospital Association, *Hospital Statistics*, 1995–1996.
7. Kelly Shriver, "What's new in subacute care?" *Modern Healthcare*, January 22, 1996.
8. National Academy of Sciences, Institute of Medicine, *Nursing Staff in Hospitals and Nursing Homes: Is It Adequate?* National Academy Press, 1996.
9. Charlene Harrington, Helen Carrillo, Susan Thollaug, and Peter Summer. *Nursing Facilities, Staffing, Residents, and Facility Deficiencies*. Department of Social and Behavior Sciences, School of Nursing, University of California at San Francisco, June 1996.
10. U.S. Department of Health and Human Services, National Institute for Occupational Safety and Health, *Workplace Use of Back Belts*, July 1994.
11. F. Bell, *Lifting Devices in Hospitals*, London: Groom Helm, 1984; and B.D. Owen, "Patient handling devices: An ergonomic approach to lifting patients," in F. Aghazadeh (Ed.), *Trends in Ergonomics/ Human Factors. V*, North-Holland: Elsevier Science Publishers, 1988.
12. Testimony of Aimee Miller, Certified Nurses' Aide, Member of Local 1199P, Service Employees International Union before the Pennsylvania House of Representatives, Committee on Labor Relations, November 19, 1992.
13. Small Business and Health Care Reform, University of Michigan School of Public Health, 1994.
14. Nursing Home Resident Rights: Has the Administration Set a Land Mine for the Landmark OBRA 1987 Nursing Home Reform Law? A Staff Report by the Subcommittee on Aging, Senate Committee on Labor and Human Resources, June 13, 1991.
15. Data from the Office of the Actuary, Health Care Financing Administration.
16. A. Garg and B. Owen, "Reducing back stress to nursing personnel: an ergonomic intervention in a nursing home," *Ergonomics*, Vol. 35, No. 11, 1992: B. Owen and A. Garg, "Assistive devices for use with patient handling tasks," in B. Das, (Ed.), *Advances in Industrial Ergonomics and Safety II*, Taylor and Francis, 1990; and R.C. Jensen and T.K. Hodous, *A Guide for Preventing Back Injuries Among Nursing Assistants Working in Nursing Homes*, U.S. Department of Health and Human Services, National Institute for Occupational Safety and Health, 1991.

17. U.S. Department of Labor, *Preventing Injury/Saving Money in Nursing Homes*, August 8, 1996.
18. G. Pascal Zachary, "Nursing homes are often hotbeds of injury for aides," *Wall Street Journal*, March 20, 1995.
19. B. Owen, K. Keene, S. Olson, and A. Garg, "An ergonomic approach to reducing back stress while carrying out patient handling tasks with a hospitalized patient" in Hagberg, Hofman, Stobel, and Westlander, Occupational Health for Health Care Workers, ECOMED, Landsberg, Germany, 1995.

10 The Occupational Hazards of Home Health Care

Elaine El-Askari and Barbara DeBaun

CONTENTS

INTRODUCTION

Health care workers who provide services in the home face a substantial risk of job-related injury and illness. These providers may be challenged by numerous environmental factors in the home, including poor sanitation, lack of ventilation, pets, rodent or insect infestations, infectious diseases, limited supplies, and unpredictable conditions and events. They are susceptible to musculoskeletal injuries due to heavy or awkward lifting, and they may even encounter the threat of violence when in the home or nearby.

This chapter is a survey of some of the occupational hazards in home health care work. It will examine the explosive growth of the homecare field and review available statistics about occupational injury and illness rates. Finally, it will focus on several key health and safety issues for home care providers—personal safety, ergonomics and back injuries, bloodborne pathogens, and tuberculosis (TB).

It should be noted that relatively little research has been done to date on home health care hazards. Many of the available injury and illness statistics are incomplete. This is clearly a field in need of further study.

0-8493-3382-2/99/$0.00+$.50
© 1999 by CRC Press LLC

TRENDS IN HOME HEALTH CARE

Health care is one of the largest industries in the United States, and home health care is the fastest growing segment of the industry. According to the Bureau of Labor Statistics (BLS), 555,400 U.S. employees worked for private home health care agencies in 1994.[1] This figure does not include employees of public agencies or hospital-based agencies. Marion Merrell Dow, a private marketing firm, conducted a survey in 1994 and found that 900,000 U.S. workers were employed in home health care overall.[2]

From 1988–93 home health care employment had an annual growth rate of 16.4%. This compared with 4.3% for total health services and 2.8% for hospitals.[3] In 1994, home health care became the second fastest growing industry segment in the entire U.S. economy. The BLS projects an increase of more than 500,000 jobs in home health care between 1994 and 2005[4] (see Figure 1).

The number of U.S. clients for home care services increased from 1.5 million in 1988 to 3.5 million in 1993. In part, this growth resulted from the expansion of Medicare coverage due to a federal district court decision *(Duggan v. Bowen)* in 1988.[5] Other contributing factors include an increasing elderly population and the lower costs for home care compared to hospital care (primarily due to the lower wages paid in home care). In addition, technological changes have made it easier to provide complex medical services in the home.

The U.S. Office of Management and Budget defines home health care services as "establishments primarily engaged in providing skilled nursing or medical care in the home, under the supervision of a physician."[6] Home health aides make up the largest proportion of home health care workers (31%). Registered nurses (RNs) constitute 20%; homemaker aides, 13%; and licensed practical (or vocational) nurses (LPNs or LVNs), 7%.[3] Other home care occupations include therapists (occupational, respiratory, physical, and speech), social workers, pathologists, and physicians. Services in the home range from basic personal care like feeding, cleaning, and bathing to more complex tasks like home infusion therapy. The skill level of home health care workers must equal or even surpass that of providers in an acute care setting. Many home care workers must be able to perform an array of advanced medical procedures in a less than ideal setting, and they often lack the resources they need.

The majority of clients cared for in the home are elderly. Many of these patients suffer from heart disease, cancer, stroke, or orthopedic problems. Many people with human immunodeficiency virus (HIV) or acquired immunodeficiency syndrome (AIDS) are being cared for in the home. A 1986 study found that approximately 17–22% of AIDS patients used home health care services at any given time.[7] A study of home care agencies in the San Francisco Bay area showed that approximately 95% of them provide care for people with AIDS.[8]

JOB INJURY AND ILLNESS IN HOME CARE

According to the BLS, there were 6588 occupational fatalities in the United States in 1994.[9] Of these, 86 fatalities occurred in health care services, including 19 in

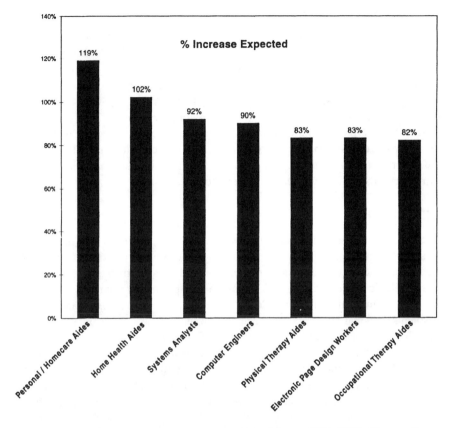

FIGURE 1   Fastest growing occupations in the United States, 1994–2005. (*Source:* Bureau of Labor Statistics [BLS], Office of Employment Projections.)

home health care. Transportation accidents caused 79% of the home health care fatalities. This can be explained by the amount of driving required between patient visits.

In the same year there were 217,817 nonfatal occupational injuries in health care services. There were 98,196 injuries in hospitals; 83,450, in nursing homes; and 18,812, in home health care.[10] Of the home health care injuries, 59% involved strains and sprains, and 49% affected the trunk (mostly the back). Common sources of injury included overexertion (39%) and interaction with the patient (36%) (see Figure 2). The home care figures probably reflect underreporting. It is often difficult for people who work in the field to report their injuries. Home care agencies usually do not have a trained employee health nurse or other occupational health staff. Therefore, home health care workers may not have access to immediate consultation, assistance, or clear reporting procedures when they sustain an injury in the home. Many of these workers are also on tight schedules, which may deter them from driving to an office or clinic out of their work area to report an injury.

OCCUPATIONAL HAZARDS

In hospitals, nursing homes, and other health care industry workplaces, employees may be exposed to numerous job hazards, including communicable diseases, chemical and biological agents, carcinogens, ionizing and nonionizing radiation, ergonomic hazards, and psychosocial risks.[11] Workers in home health care are exposed to many of these same hazards. In addition, they work in a unique environment where they can also be faced with a range of other problems. This chapter will address a few typical home care hazards. However, the authors stress that there is a scarcity of research on occupational hazards in home health care. Therefore, we will describe relevant information from the literature and add examples from our own experience with home health care employees.

THE SETTING

To understand the hazards faced by home care workers, one must take into account the unique conditions found in the home care setting. Smith[12] identified three distinct types of home care work environments: rural, suburban, and urban inner city. According to Smith, common physical risk factors in all types of environments include pets and other animals; driving and parking; home structure, maintenance, and hygiene; personal safety and vulnerability; weather; and time of day. These and other factors result in a lack of worker control over the environment.

Smith conducted interviews with 29 home health care workers, and observed them as they went about their work. Workers in rural areas related several stories about the unusual risks they faced. For example, a nurse was walking up to a house when a pig "...ran towards me and wham, it bit me. I was just so stunned. It was hard to live that one down with my fellow workers and the ER workers who treated me." Another nurse was confronted by a gaggle of geese "hissing and flapping their wings." Other workers described the hazards of driving on unpaved country roads full of potholes in bad weather. Some had experienced car trouble with no one nearby to help. (See Figure 2.)

In a suburban area, one nurse described working in homes where people let their pets "do their business" all over the house. Another worker reported coming out of a suburban home "with flea bites, and covered in dog or cat hair."

Many of the problems described by workers in the inner city involved personal safety. Workers described coming in contact with pit bulls (extremely aggressive dogs). Others said that they had to work around loaded guns and other kinds of weapons. One worker described some of the hazards associated with housing projects: "The high-rise projects can be very dangerous. Stairwells are especially bad spots (with a lot of drug use and dealing). Elevators can also be a problem. You could call the elevator and walk in on a drug deal. In some buildings, the elevators are controlled by the dealers."

PERSONAL SAFETY

Violence on the job has emerged as an important safety issue in many occupations today. The BLS reports that in 1994 there were approximately 20,000 nonfatal violent

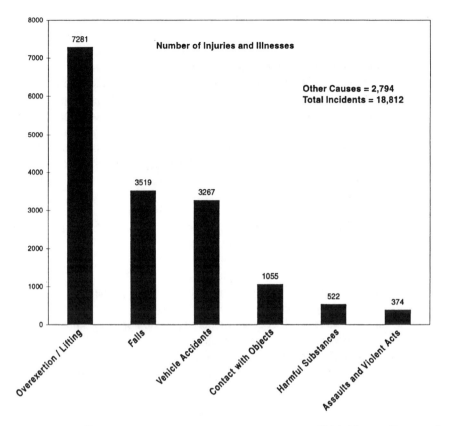

FIGURE 2 Leading causes of injury and illness in home care, 1994. (*Source:* Bureau of Labor Statistics.)

incidents in the nation's workplaces (counting only those that resulted in lost work-days). Women were the victims in nearly 60% of these incidents. The most common nonfatal violent acts (about 40%) involved "hitting, kicking, and beating."[13]

Service workers are at especially high risk of nonfatal assault. Of the service workers who are assaulted, nursing aides and orderlies account for more than half. These and other workers in health and residential care occupations (e.g., social workers) are assaulted primarily by clients who resist their help. There have been no studies to date that examined the risk of assault in home care specifically.

Federal OSHA issued its "Guidelines for Preventing Workplace Violence Among Health Care and Social Service Workers" in March 1996.[14] OSHA identified home care workers as a group at risk for workplace violence because they "work alone ... may have to work late night or early morning hours ... often work in high-crime areas ... and work in homes, where there is extensive contact with the public."[15]

The risks to personal safety in home health care have not been adequately studied, and statistics on the prevalence of the problem may be unreliable. This is another area where underreporting may be significant. Some researchers have suggested that home care workers (like other health care workers) often accept assaultive behavior as part of the job. In addition, formal incident reports may be filed only for assaults that require medical treatment. Often no reports are filed for verbal threats, sexual harassment, or various other types of assault.[16]

One study sought to determine whether the quality of care is affected when home visits present threats to personal safety.[17] Nearly 100 home care employers and administrators were asked about their perception of risk and response to it. Among the factors that respondents associated with risk were geographic location, high incidence of crime, inappropriate patient or caregiver behavior, and evening assignments. Respondents emphasized that visits are now made to the home 24 hours a day. Approximately 66% of respondents said that they leave a situation "as soon as possible" if they perceive it to involve high risk. A larger study is now in progress by the same authors.

Under U.S. health and safety law, the employer is responsible for maintaining a safe and healthful workplace. OSHA has outlined several steps that employers can take to prevent violence in the home care setting:

- Provide safety education for employees.
- Establish a communication system (such as cellular phones).
- Utilize a "buddy system" or security escorts.
- Establish policies to discourage robbery (e.g., do not carry a purse).
- Assure proper maintenance of vehicles.
- Provide handheld alarms to employees.
- Set up procedures for employees and the employer to follow when violence does occur.

(Also see Figure 3.[17a])

ERGONOMICS AND BACK INJURIES

According to BLS figures for 1994, overexertion was the most common cause of lost time occupational injury in the private sector U.S. workforce.[13] Nursing homes and scheduled airline or air courier services led all other industries in overexertion incidents. The same year, the BLS rated the ten occupations with the most lost time injuries and illnesses of all types, and found that nursing aides and orderlies ranked third. (Truck drivers were first, and nonconstruction laborers were second.)

The BLS found that sprains and strains were, by far, the leading type of work-related injury in every major U.S. industry. The trunk, including the back, was the body part most affected. Overexertion while maneuvering objects led all other disabling events and was cited in 16–33% of the cases in every industry.

One study examined low-back injuries in health care during 1984–1986.[18] Incident reports were collected for nursing assistants in a large hospital and home health aides in two large home care agencies. The researchers found the rate of back injury to be much higher among the home health aides (15.4 per 100 full-time equivalent [FTE] workers) than among the hospital nursing assistants (5.9 per 100 FTE workers).

✔

☐   Are workers briefed about the areas where they will be working (gang colors, neighborhood culture, language, drug activity, etc.)?

☐   Are workers notified of past violent acts by particular clients?

☐   Are workers given maps and good directions covering the areas where they will be working?

☐   Do workers avoid carrying valuables into the home (purses, etc.)?

☐   Do workers have two-way radios, pagers, or cellular phones for communication in an emergency?

☐   Do workers carry personal alarm devices?

☐   Is safe parking readily available near clients' homes?

☐   Are vehicles kept in good working order to minimize breakdowns?

☐   Are door and window locks in vehicles easily controlled by the driver?

☐   Are escorts or "buddies" provided when people work in potentially dangerous situations?

☐   Is training provided to all workers on personal safety and violence prevention?

☐   Are incident reporting and follow-up procedures in place?

FIGURE 3   Personal safety tips for home health care workers. (*Source:* Adapted from *Violence on the Job: A Guidebook for Labor Management,* Labor Occupational Health Program [LOHP], University of California, Berkeley, 1997.)

The study showed that 40% of the back injuries in both groups involved activities at the patient's bedside. About 66% of the back injuries occurred during planned patient care activities such as transferring a patient to a wheelchair. For home care, other activities at the time of injury included helping a patient in or out of a bed, chair, or tub; catching a patient during a fall; turning a patient; stooping over a patient in bed; and helping a patient on or off a toilet. No lifting equipment was used in 75% of the incidents involving nursing assistants and 80% of the incidents among home health aides. Home health aides were working alone in 88% of the incidents, and nursing assistants were working alone in 39% of the incidents.

One of the problems in the home care setting is that the worker has to adapt to an unpredictable environment, which differs from one home to the next. There is no opportunity to engineer safety problems out of the home as can be done with fixed workplaces. Adaptation is even more difficult because it must typically be done by a single employee working alone, for many clients in many different homes. For example, the "lifting team" concept developed for the acute care setting cannot be applied to home care. Charney and co-workers[19] describe the lifting team concept as a method that incorporates proper body mechanics, personal protective equipment (such as transfer belts), mechanical lifting devices, and a two-person team. Candidates for the lifting team are screened for flexibility and strength. Those selected for the team are trained on proper lifting techniques, how to use mechanical lifting devices, and how to coordinate the lift. Although this full concept is not applicable to home care because most employees work by themselves, it is at least important to provide mechanical lifting devices and to train home care workers in how to use them.

Owen[20] recommends such an approach to reduce back injuries in the home care setting. The patient's needs should be assessed; and the proper techniques should be selected to handle, lift, and move the patient safely. These techniques should then be clearly communicated to the staff who will be doing the work. This communication should be included in new employee orientation programs, and there should be frequent in-service training. Owen stresses that "assistive devices or mechanical lifts" should be used to lift or transfer patients and to move patients in bed. Devices used in the home will have to be different from the heavy equipment often used in hospitals and other institutions. Devices should be easy to maneuver and carry into the home. Owen suggests transfer belts, transfer boards, portable commodes, chairs and cushions that lift, and lift poles that a patient can use to stand up independently.

INFECTIOUS DISEASES

Home health care workers must be protected against communicable diseases at all times. A range of personal protective equipment should be readily available, including gloves, face shields, impervious gowns, and cardiopulmonary resuscitation (CPR) masks. Respirators may be needed when working with TB patients. Equipment should also be available for protection against bloodborne diseases, including safe needle devices that protect against accidental needlestick injury, good sharps disposal containers, and access to immediate treatment when there is a significant blood exposure.

BLOODBORNE PATHOGENS

Occupational exposure to bloodborne pathogens continues to be a threat to health care workers both in institutions and in the home. Today there is a great deal of concern about exposure to the hepatitis B virus (HBV) and HIV, both of which are transmitted through blood contact. Many more health care workers become infected with HBV than with HIV, but HIV is still considered a significant risk.

The Centers for Disease Control and Prevention (CDC) reported on 51 documented cases of occupational transmission of HIV to health care workers. (There are 106 additional cases where occupational transmission is considered a possibility but has not been documented.) Percutaneous exposure was the source in the majority

(44) of the 51 documented cases. (It is generally believed that CDC figures represent considerably less than the actual number of cases, due to underreporting.)

The occupational risk of exposure to bloodborne pathogens in the home care setting has not been the focus of a great deal of investigation or research. For example, the number of needlestick injuries in home health care is unknown, although it has been estimated that 800,000 needlesticks occur each year in acute care institutions in the United States.[22] Activities that pose a risk of blood exposure to home health care workers include intravenous (IV) infusions, vascular access, IV nutritional support, wound care, and respiratory therapy. The risk of a needlestick injury in this setting may be increased due to the less controlled environment (poor lighting, clutter, cramped quarters, pets, etc.). One home health care nurse reported getting a needlestick because a client's cat jumped on the bed while she was drawing blood. Also, proper equipment may not be available in home care.

In 1994, the CDC conducted an epidemiological survey involving bloodborne pathogens in the home. CDC summarized eight reported cases of home caregivers who became infected with HIV. In four of these cases, the caregivers had significant direct contact with an infected person's body secretions or excretions.[23]

Askari and Lipscomb[24] conducted a 1994 study that compared needlestick injury rates between home care nurses and hospital nurses. An anonymous questionnaire, completed by 76 Northern California home care nurses, sought to determine the number of reported and unreported needlestick injuries in the group. Results showed that a substantial number of needlestick injuries occur to home care nurses and that the majority of them go unreported. The rate of *reported* needlestick injuries among home care nurses was 50% of the rate among the acute care hospital nurses. However, the home care rate may be much greater than these figures indicate because of underreporting. The survey found that 87% of the needlestick incidents in home care were unreported.

UNIVERSAL PRECAUTIONS

Most home health care agencies are required to comply with the federal OSHA bloodborne pathogens standard.[24a] The standard mandates a set of protective measures termed "Universal Precautions." This approach is based on treating *all* blood and body fluids as though they are infectious. Employers must provide, at no cost, equipment such as gloves, gowns, masks, mouthpieces, resuscitation bags, and sharps disposal containers, as well as employee training.

Askari[25] conducted a 1995 study to assess compliance with universal precautions in home health care. A questionnaire was completed by 214 home health care nurses in northern California. Results showed that compliance rates were highest for glove use and proper sharps disposal; and lowest for wearing fluid-resistant gowns, face shields, and eye protection.

When asked about the obstacles encountered to use of universal precautions in the home, 66% of the respondents said that they had no face shields available, 38% had no gowns, and 28% had no eye protection. The respondents identified availability of equipment as the major factor determining compliance with universal precautions.

Approximately 47% of the respondents also indicated that "feeling comfortable using protective equipment" affected their decision to use it.

Respondents said that they had problems with sharps containers, on the average, about 25% of the time. Problems included needle apparatus getting caught in the container (37%); no container being nearby (32%); container being full (28%); and caregiver being forced to use a bottle or can in place of a real sharps container (27%).

In response to this identified need, the National Institute for Occupational Safety and Health (NIOSH) has awarded new funding to the Training for Development of Innovative Control Technology Project (TDICT) in San Francisco and the Labor Occupational Health Program (LOHP) at the University of California, Berkeley to design a prototype safety device specifically to reduce exposure to bloodborne pathogens in home care. Home care workers will be fully involved throughout the design process.

In 1993–1994, LOHP received a grant from federal OSHA to train California home care workers on the bloodborne pathogens standard. The training was conducted with the Service Employees International Union (SEIU). LOHP and SEIU trained 285 home care workers. Over 75% were aides and attendants (not nurses). Approximately 81% were women, and 67% were people of color. Anecdotal remarks made by participants during the training revealed that most of these workers had never received any information from their employers about bloodborne pathogens or how to protect themselves on the job. In addition, the majority had to purchase their own protective equipment (gloves, aprons, etc.). They were among the lowest paid workers in the entire health care industry, and were clearly not aware of protections mandated by the OSHA standard.

TUBERCULOSIS

Since 1984, the annual number of new TB cases in the United States has increased by 18%—from 22,255 cases in 1984 to 26,283 cases in 1991.[26] A disease that was once thought to be under control has reemerged. Since 1993, however, there has been a slight decline in the annual number of new cases. The CDC reports that 24% of TB cases in the United States occur in people age 65 and older. This has important implications for home care workers, who see many elderly patients.

Studies have shown that health care workers have a significant risk of occupationally acquired TB. Surveillance studies show that some groups of workers who care for TB patients become infected at an annual rate of 5–10% (purified protein derivative [PPD] conversion rate).[27] However, it must be stressed that *none* of these studies have surveyed workers in home care.

Less than 1 page of the CDC 132-page "Guidelines for Preventing the Transmission of Mycobacterium Tuberculosis in Health Care Facilities" is devoted to home health care.[28] It is critical that more study be done and that home health care agencies establish comprehensive TB training, education, counseling, and screening programs. There should be provisions for identifying workers and clients who have active TB, baseline two-step PPD skin testing, and follow-up skin testing at intervals appropriate to the degree of risk.

CONCLUSIONS

Home care is one of the fastest growing industries in the United States, and home care aides are projected to be the fastest growing occupation in the industry. These aides are mainly middle-aged women who have not graduated from high school.[29] In the San Francisco Bay area, they are primarily women of color. They are some of the lowest paid workers in health care, and have the least knowledge about the risks they face on the job. They receive little training. Although they are the fastest growing sector of the workforce, they have very limited power to make changes in their work environment.

Turnover rates are high. Although their interests might best be served by joining a union that can bargain for better working conditions, only about 10% of U.S. home care workers are organized to date.

Worker safety, client safety, quality of care, and client autonomy are intertwined tightly in the home care setting. There is little documentation or research on the hazards faced by home care workers. It is clear, however, that they do face unique health and safety challenges, which are increased due to an unpredictable work environment as well as poor training and equipment. The home environment is often beyond the control of both the worker and the client. Crime in the neighborhood, unsafe or poorly maintained buildings, inadequate lighting, and poor sanitation can impact them both.

Following are some problems that have been addressed in this chapter:

- Many home care workers are not given adequate training on how to safely lift, transfer, or reposition a patient. Many are not provided assistive devices that can help them perform these tasks in a proper ergonomic manner. The result can be potential worker and client injuries.
- Few steps are being taken to analyze assaults and violent acts against home care workers. Also there is often a lack of education and training, safety procedures, and security devices to protect these workers.
- Many home care workers are not provided adequate training on infection control policies and procedures (especially regarding bloodborne pathogens and TB). This puts both the patient and worker at risk of infectious disease.
- Many home care nurses are performing technologically advanced medical procedures that may entail high risk. For example, IV catheter insertion and care, wound care, and tracheotomy care may now be done in the home. The hazards of performing such procedures in the unique environment of home care have not been adequately studied. Safety devices designed specifically for the home care setting may need to be developed.
- OSHA does not inspect home workplaces. Thus there is no objective way to assess compliance with its regulations, such as the hazard communication and the bloodborne pathogens standards.
- Sometimes no personal protective equipment is supplied to home care workers, or they are expected to supply it themselves. Equipment that is provided may be inadequate.

As employers, home care agencies should establish procedures and policies to address the hazards faced by their workers. Workers' input should be solicited. A safer work environment should serve to foster a better working relationship between clients and those who care for them.

REFERENCES

1. U.S. Department of Labor, Bureau of Labor Statistics (BLS). *Occup. Outlook Q*, Fall 1995.
2. Service Employees International Union, Research Department. The home care industry. unpublished report, 1995.
3. Freeman L. Home-sweet-home health care. *Monthly Labor Rev*, March 3–11, 1995.
4. U.S. Department of Labor, Bureau of Labor Statistics (BLS), Office of Employment Projections. Fastest growing occupations 1994–2005. November 1995.
5. *Duggan v. Bowen*. 691 F. Supp. 1487 (D.D.C. 1988).
6. U.S. Office of Management and Budget. Standard Industrial Classification Manual, 1987.
7. Lusby G. Martin JP, Schietinger H. Infection control at home: a guideline for caregivers to follow. *Am J Hospice Care*, March/April 1986.
8. White M, Smith W. Home health care: infection control issues. *Am J Infection Control*, 21(3), 1993.
9. U.S. Department of Labor, Bureau of Labor Statistics (BLS). Fatal occupational injuries by industry and event or exposure, 1994.
10. U.S. Department of Labor, Bureau of Labor Statistics (BLS). Survey of occupational injuries and illnesses, 1994.
11. Sterling D. Overview of health and safety in the health care environment. *Essentials of Modern Hospital Safety*, Vol. 3, Chapter 1, Lewis Publishers, Boca Raton, FL, 1994.
12. Smith W. Occupational risk perception in home health care workers. Doctoral dissertation, University of Michigan, UMI No. 9542205, 1995.
13. U.S. Department of Labor, Bureau of Labor Statistics (BLS). Characteristics of injuries and illnesses resulting in absences from work, 1994.
14. U.S. Department of Labor, Occupational Safety and Health Administration. Guidelines for preventing workplace violence among health care and social service workers, 1996.
15. U.S. Department of Labor, Program Highlights. Protecting community workers against violence. Fact Sheet No. OSHA 96-53, 1996.
16. Lipscomb J, Love C. Violence toward health care workers—an emerging occupational hazard. *AAOHN J*, 40(5), May 1992.
17. Kendra A, Weiker A, et al. Safety concerns affecting delivery of home health care. *Public Health Nursing*, 13(2), April 1996.
17a. Violence on the Job: A Guidebook for Labor Management. Labor Occupational Health Program (LOHP), University of California, Berkeley, 1997.
18. Myers A, Jensen R. et al. Low back injuries among home health aides compared with hospital nursing aides. *Home Health Care Q*, 14(2/3), 1993.
19. Charney W., Zimmerman K, Walara M. The lifting team. *AAOHN J*, 39, May 1991.
20. Owen B. Back injuries in the home health care setting, *Home HealthCare Consultant*, 3, May/June 1996.
21. U.S. Centers for Disease Control and Prevention (CDC). HIV/AIDS Surveillance Report, June 1996.

22. Jagger F., Pearson RD. Universal Precautions: still missing the point on needlesticks. *Infect Control Hosp Epidemiol,* 12(4), 1991.
23. U.S. Centers for Disease Control and Prevention (CDC). Epidemiologic notes and reports, human immunodeficiency virus transmission in household settings—United States. *MMWR,* 19, 1994.
24. Askari E, Lipscomb J. Bloodborne pathogens—risks in the home health care setting. *AAOHN J,* submitted for publication.
24a. Code of Federal Regulation (CFR) 29, Part 1910. 1030.
25. Askari E. Compliance with universal precautions in home care, unpublished study, 1995.
26. U.S. Centers for Disease Control and Prevention (CDC).
27. Markowitz S. Epidemiology of tuberculosis among health care workers. *Occup Med: State Art Rev,* 9(4), October–December 1994.
28. U.S. Centers for Disease Control and Prevention (CDC). Guidelines for preventing the transmission of mycobacterium tuberculosis in health-care facilities.
29. U.S. Bureau of the Census. Current Population Survey, 1994.

Appendix 1
Findings of Minnesota Nurses Association Research Project on Occupational Injury/Illness in Minnesota Between 1990 and 1994*

Elizabeth Shogren and Andrew Calkins

CONTENTS

BACKGROUND

Following congressional hearings in 1993, Congress directed the Department of Health and Human Services to commission a study concerning the state of staffing of nursing personnel in hospitals and nursing homes in the United States.

The Institute of Medicine (IOM), National Academy of Sciences, conducted a comprehensive study to determine whether, and to what extent, there is a need for an increase in the number of nurses in hospitals and nursing homes to promote the quality of patient care and reduce the incidence among nurses of work-related injuries and stress.

The IOM established a committee to receive testimony and examine and evaluate available data on this topic. Minnesota Nurses Association was one of dozens of

---

* Reprinted with permission from the Minnesota Nurses Association, which provided funding for the study (copyright 1997).

organizations that prepared written testimony. In January of 1995, Minnesota Nurses Association was invited to present oral testimony relative to its concerns. In particular, the IOM expressed an interest in Minnesota Nurses Association's concern about increased injuries associated with restructuring of the health care system in Minnesota.

In presenting testimony to the IOM, Minnesota Nurses Association put forth a hypothesis that cost containment measures such as diversion of less acute patients to outpatient settings; decreased length of stay; layoffs and work redesign, which had been implemented in response to increased managed care; prospective reimbursement capitation; and deep discounting had produced workplace conditions likely to cause increased injury or illness.

In response to that testimony, Minnesota Nurses Association was asked to produce data that supported its concerns. Minnesota Nurses Association initiated a data collection project in March 1995. Data were collected from all facilities where Minnesota Nurses Association was the certified bargaining representative. Not all those data are included in this report.

What follows is a summary of data collected from 11 private hospitals located in the Minneapolis and St. Paul (Twin City) Metropolitan area for the years 1990, 1992, and 1994. Those hospitals are

Abbott Northwestern Hospital	Minneapolis, MN
Bethesda Lutheran Hospital	St. Paul, MN
Children's Health Care—St. Paul	St. Paul, MN
Children's Health Care—Minneapolis[a]	Minneapolis, MN
Fairview Riverside Medical Center	Minneapolis, MN
Fairview Southdale Hospital	Edina, MN
Mercy Medical Center	Coon Rapids, MN
Methodist Hospital	St. Louis Park, MN
North Memorial Hospital	Robbinsdale, MN
Phillips Eye Institute[a]	Minneapolis, MN
St. John's Hospital	Maplewood, MN
St. Joseph's Hospital	St. Paul, MN
United Hospital	St. Paul, MN

Note: All hospitals with the exception of Methodist Hospital participated in multi-employer bargaining for a 1995–1998 contract with Minnesota Nurses Association. Methodist Hospital bargained concurrently, but separately for a 1995–1998 contract with Minnesota Nurses Association.

[a] Injury and illness data for these facilities are not complete and therefore are not included in this report. However, these hospitals are included in other data.

METHOD

In its capacity as a certified bargaining representative, Minnesota Nurses Association has access to the complete Occupational Safety and Health Administration (OSHA) 200 Logs at each facility. The OSHA 200 Logs are documents required under the federal and Minnesota State Occupational Safety and Health Acts. Employers are required to record occupational injuries and illnesses that meet specified criteria.

OSHA recordable incidents are employee injuries or illnesses related to performance of work tasks or work exposure that:

- Causes lost time from work past the date of injury
- Causes the employee to have a restricted work assignment or transfer to another job
- Requires treatment beyond first aid
- Results in the loss of consciousness or death

Minnesota Nurses Association accessed and collected the OSHA 200 Logs from each facility for the years 1990, 1992, and 1994.

Data were retrieved for the following job classifications:

1. Registered nurses (RNs)
2. Licensed practical nurses (LPNs)
3. Nursing assistants
4. Other unlicensed personnel (primarily technical personnel, e.g., dialysis technicians, phlebotomists, and x-ray technicians)
5. Other professionals (e.g., pharmacists, physical therapists)

Data retrieved included the following categories of information. Data were entered using the coding procedure for OSHA logs developed by the Department of Labor:

1. Date of injury
2. Occupation (as categorized earlier)
3. Description of injury
4. Body part affected
5. Nature of injury
6. Source of injury
7. Event, if identifiable
8. Lost workdays
9. Restricted workdays

The only change in what constituted a recordable injury or illness during 1990–1994 was a change in the criteria for what constituted a recordable needlestick or sharps injury for 1994. The impact on the data for 1994 would be a probable decrease in the number of recordable injuries for 1994.

Other sources of data were

1. The Center of Healthcare Industry Performance Studies (CHIPS) reports in the 1993 Almanac of Financial and Operating Indicators
2. Minnesota Department of Health

3. The Annual Actuarial Valuation Report for Twin City Hospital, Minnesota Nurses Association Pension Plan for the years 1990–1994 compiled by William M. Mercer, Inc.

4. Data generated by Metropolitan Health Care Council (MHC) and Minnesota Nurses Association during the course of bargaining the 1995–1998 contract agreement(s)

In developing this project, there was an attempt to rely on data that were relatively easy to obtain and that would not be subject to questions of accuracy.

The data on injury and illness were provided by the employers in all cases. Each of the facilities had occupational health services under the direction of qualified occupational health professionals who were knowledgeable about OSHA requirements and criteria for recordable injuries.

The data were extrapolated from the OSHA 200 Logs by one person. In any situation where a question arose concerning an entry, the hospital was contacted for clarification. Questions about appropriate coding were referred to the Minnesota Department of Labor and Industry. Other data were generated from public records, industry publications, pension records, or information provided by the employers during negotiation.

FINDINGS

1. RN bargaining unit positions (staff nurse) decreased from 1990 to 1994:

Year	RN Bargaining Unit Positions
1990	7367
1992	6755 (9% reduction from 1990)
1994	6712 (9.8% reduction from 1990)

2. Overall, RN positions (bargaining unit and nonbargaining unit) in the identified hospitals decreased from 1990 to 1992 and despite a slight recovery in 1994, remained significantly below 1990:

Year	RN Positions
1990	8951
1992	7836 (14.2% reduction from 1990)
1994	8036 (10.2% reduction from 1990)

Source: The Annual Actuarial Valuation Report for Twin City Hospital, Minnesota Nurses Association Pension Plan for the years 1990–1994 complied by William M. Mercer, Inc.

3. The aggregate numbers of recordable injuries for all classifications of employees rose 64.1% from 1990 to 1992; and 70% from 1990 to 1994:

Year	Recordable Injuries for All Classifications of Employees
1990	1138
1992	1867 (64.1% reduction from 1990)
1994	1923 (70% reduction from 1990)

Source: OSHA 200 Logs complied by identified hospitals.

4. Each classification of employees had increases in the number of injuries:

	1990–1992		Increase (%)	1990–1994		Increase (%)
RNs	569	921	61.8	569	940	65.2
LPNs	106	142	33.9	106	159	50.0
Nursing assistants	196	312	59.7	196	295	51.0
Other UAP	205	359	75.1	205	444	116.5
Other professionals	62	133	114.5	62	105	85.4

Source: OSHA 200 Logs complied by identified hospitals.

5.* • The percentage of recordable injuries and illnesses for RNs in 1990 was 6.35%.
   • The percentage increase in recordable injuries and illnesses for RNs from 1990 to 1992 was 11.75%.
   • The percentage increase in recordable injuries and illnesses for RNs from 1990 to 1994 was 11.57%.
   • At this time, there are inadequate data to evaluate this for all categories of employees or to calculate a rate of injury as that term is defined by OSHA.

6.* • All employers, except one, showed increases in three or more categories of employees from 1990 to 1992 and 1990 to 1994.
   • Two employers showed increases in all categories of employees from 1990 to 1992 and 1990 to 1994.
   • One employer showed a decrease in most categories from 1990 to 1992 and 1990 to 1994. (This hospital's recordable injuries and illnesses for 1990 were very high compared to the other hospitals.)

---

* *Source*: OSHA 200 Logs complied by identified hospitals.

7. The average percentage of RN full-time equivalents (FTEs) to total FTEs remained relatively constant from 1990 to 1994.

Total RN FTEs to Total FTEs

	1990	1991	1992	1993	1994
Minnesota State average for contract hospitals	27.9	27.8	27.3	27.3	27.4

Source: Minnesota Department of Health, Health Economics Program.

8. The number of RN FTEs to patient care days increased.

Total RN FTEs to 100 Patient Care Days

	1990	1991	1992	1993	1994
Average for Hospital identified	0.487	0.504	0.531	0.567	0.563
	(1013 h)	(1048 h)	(1104 h)	(1179 h)	(1171 h)
RN hours per patient day	10.13	10.48	11.04	11.79	11.71

Note: 1990–1991, 3.15% increase; 1991–1992, 5.34% increase; 1992–1993, 6.36% increase; 1993–1994, 0.71% increase; 1990–1994, 15.61% aggregate increase.

Source: Minnesota Department of Health, Health Economics Program.

9. The average length of stay (LOS) decreased from 1991 to 1994.

Year	Average LOS
1990	Data not available
1991	5.9 days
1992	5.7 days
1993	5.3 days
1994	5.2 days

Source: Metropolitan Health Care Council.

10a. The number of outpatient visits for identified hospitals increased:

Year	Number of Outpatient Visits (Thousands)	Increase (%)
1990	Data not available	
1991	902.2	
1992	918.8	1.8
1993	916.0	0.3
1994	931.9	1.74
1991–1994		3.29

10b. The number of adjusted patient days decreased:

Year	Adjusted Patient Days (Thousands)	Decrease (%)
1990	Data not available	
1991	1292.3	
1992	1233.6	4.5
1993	1136.0	7.9
1994	1093.8	3.7
1991–1994		18.15

Source: Metropolitan Health Care Council.

11. The 1989–1994 cost containment activities include:

Layoffs

• 5 Layoffs at Fairview Riverside Medical Center	1989–1992
• 8 Layoffs at Fairview Riverside Medical Center	1992–1994
• 1 Layoff at Methodist	1993
• 2 Layoffs at Fairview Southdale	1993–1994
• 2 Layoffs at Minneapolis Children's Hospital	1991 and 1994
• 2 Layoffs at St. Paul Children's Hospital	1991 and 1994
• 1 Layoff at Mercy Hospital	1993
• 2 Layoffs at Abbott Northwestern	1991–1994
• 3 Layoffs at Health East	1992–1994

Note: Each layoff affected varying numbers of RNs ranging from 5 to 100 RNs plus other classifications of employees.

Consultants/Redesign

American Practices Management	Fairview Riverside Medical Center	1989
American Practices Management	Mercy Hospital	1992–1993
Anderson	United Hospital	1992

Closure Impact on Other System Hospitals

1992	Metropolitan-Mount Sinai closure impacts United and Mercy Hospitals through relocation of employees services and creates new entity: Phillips Eye Institute.
1994	Closure of Divine Redeemer impacts Bethesda Hospital, St. John's Hospital, and St. Joseph's Hospital by reduction of beds and services that precipitates displacement of nurses in those facilities to absorb nurses affected by closure of Divine Redeemer.

Source: Minnesota Nurses Association.

12. The most prevalent injuries and illnesses for each year are as follows:

1990 Nature of Injuries/Illnesses	Number of Injuries/Illnesses
1. Traumatic injuries to muscles, tendons, ligaments, joints, etc.	506
2. Other traumatic injuries and disorders	230
3. Blood and body fluid exposures (primarily needlestick and sharps)	196
4. Surface wounds and abrasions	75
5. Open wounds	63
6. Disorders of skin and subcutaneous tissue	26

1992 Nature of Injuries/Illnesses	Number of Injuries/Illnesses
1. Traumatic injuries to muscles, tendons, ligaments, joints, etc.	721
2. Blood and body fluid exposures (primarily needlestick and sharps)	654
3. Other traumatic injuries and disorders	225
4. Surface wounds and abrasions	107
5. Disorders of skin and subcutaneous tissue	105
6. Open wounds	61

1994 Nature of Injuries/Illnesses	Number of Injuries/Illnesses
1. Traumatic injuries to muscles, tendons, ligaments, joints, etc.	639
2. Blood and body fluid exposures (primarily needlestick and sharps)	599
3. Other traumatic injuries and disorders	319
4. Surface wounds and abrasions	156
5. Disorders of skin and subcutaneous tissue	107
6. Other infectious and parasitic diseases	64

1992–1994 Nature of Injuries/Illnesses	Number of Injuries/Illnesses
1. Traumatic injuries to muscles, tendons, ligaments, joints, etc.	1866
2. Blood and body fluid exposures (primarily needlestick and sharps)	1449
3. Other traumatic injuries and disorders	774
4. Surface wounds and abrasions	338
5. Disorders of skin and subcutaneous tissue	238
6. Open wounds	169

Source: OSHA 200 Logs complied by identified hospitals.

DISCUSSION

The undisputed shift of less acute patients from inpatient to outpatient settings would obviously result in an increase in average acuity of patients who remain in the hospital. Acuity is a concept that reflects the time it takes to provide required care to a patient. In general, patients who are more ill or require more lengthy or complex care have a higher acuity. The data demonstrate a 15.61% aggregate increase in RN hours per patient day while the percentage of RN FTEs to total FTEs remains stable from 1990 to 1993, may also be a strong indicator that patient acuity increased over this time frame.

At the same time, LOS for those remaining patients decreased from 5.9 to 5.2 days. Together, increased average acuity and decreased length of stay results in more concentrated work; that is, more work must be done in less time.

Concurrently, other events occurred. Frequent layoffs, changes in delivery models and work processes, and hospital closures and mergers created additional stresses and an unstable workplace. Any one of these: more work, faster work, or unstable work environment would be likely to create a potential for increased injuries and illnesses.

CONCLUSIONS

The data collected demonstrate a significant and sustained increase in the numbers of recordable injuries for all categories of health care workers from 1990 to 1994.

The diversion of less acute patients to outpatient services and the decreased length of stay for inpatient clients are critical components of articulated strategies and goals of insurers and health maintenance organizations (HMOs) in Minnesota to reduce costs. Public payors, that is, Medicare and Medicaid, had payment schedules based on diagnosis-related groups (DRGs) in an attempt to reduce costs. These, and other cost containment strategies, in turn, pressured the hospitals to implement a variety of cost-saving measures including layoffs of staff, mergers, closures, reorganization of services, and restructuring of the care delivery models.

The data strongly suggest a causal relationship between increased concentration of work and increased injury and illness. This relationship needs to be examined further to isolate the identified variables. In addition, data should be collected and analyzed on an ongoing basis to assess the impact of cost-saving strategies, on workplace injury and illness, or the effectiveness of efforts to control and reduce workplace injury and illness in this type of climate in Minnesota and elsewhere.

REFERENCE

Rogers, B. Nursing injury, stress and nursing care. In *Nursing Staff in Hospitals and Nursing Homes: Is It Adequate?* Edited by G. Wunderlich, et al. Washington, D.C., National Academy Press, 1996, pp. 503–532.

Appendix 2
Hospital Injury
Data Charts

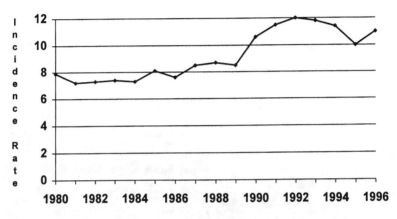

Hospital Incidence Rate for Total Cases

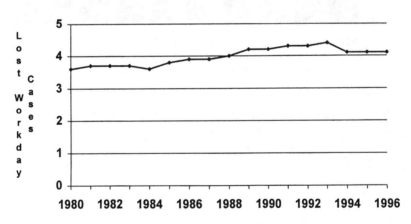

Hospital Incidence Rates for Lost Workday Cases

0-8493-3382-2/99/$0.00+$.50
© 1999 by CRC Press LLC

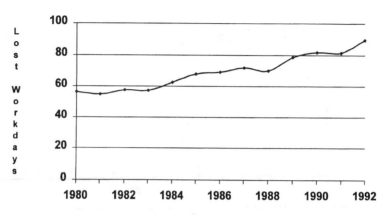

Hospital Incidence Rates for Lost Workdays

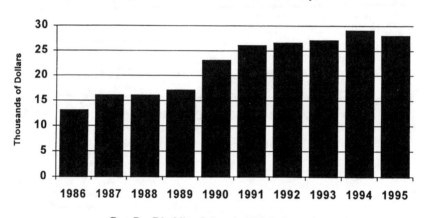

Cost Per Disabling Injury in U.S. Industries

Health Care Occupational Injuries and Illnesses Due to Strains and Sprains
Percentage of Total Workers

Registered Nurses	Licensed Practical Nurses	Nurses' Assistants, Orderlies, and Attendants
62.1	64.7	65.5

Source: Bureau of Labor and Statistics (BLS), 1993.

Health Care Occupational Injuries and Illnesses Due to Patients
Percentage of Total Workers

Registered Nurses	Licensed Practical Nurses	Nurses' Assistants, Orderlies, and Attendants
45.4	48.7	61.3

Source: BLS, 1993.

Hospital Incidence Rates

Year	Total Cases	Lost Workday Cases	Lost Workdays
1996	11.0	4.1	—
1995	10.0	4.1	—
1994	11.4	4.1	—
1993	11.8	4.4	—
1992	12.0	4.3	89.5
1991	11.5	4.3	81.5
1990	10.6	4.2	81.6
1989	8.5	4.2	78.7
1988	8.7	4.0	70.0
1987	8.5	3.9	71.9
1986	7.6	3.9	68.9
1985	8.1	3.8	67.9
1984	7.3	3.6	62.6
1983	7.4	3.7	57.2
1982	7.3	3.7	57.5
1981	7.2	3.7	54.7
1980	7.9	3.6	56.2

Note: Standard Industry Code (SIC) 806.

Nursing and Personal Care Facilities Incidence Rates

Year	Total Cases	Lost Workday Cases	Lost Workdays
1996	16.5	8.3	—
1995	18.2	8.8	—
1994	16.8	8.4	—
1993	17.3	8.9	—
1992	18.6	9.3	212.5
1991	15.3	8.7	186.9
1990	15.6	9.1	190.3
1989	15.5	8.8	181.8
1988	15.0	8.7	180.6
1987	14.2	8.0	158.9
1986	13.5	7.7	129.0
1985	13.3	7.3	136.4
1984	11.6	6.5	121.3
1983	11.0	6.0	98.2
1982	10.1	5.7	95.1
1981	10.3	5.6	90.7
1980	10.7	5.6	83.5

Note: SIC 805.

Index